Advanced Space System Concepts and Technologies: 2010–2030+

Advanced Space System Concepts and Technologies: 2010–2030+

Ivan Bekey

The Aerospace Press • El Segundo, California

American Institute of Aeronautics and Astronautics, Inc. • Reston, Virginia

The Aerospace Press
2350 E. El Segundo Boulevard
El Segundo, California 90245-4691

American Institute of Aeronautics and Astronautics, Inc.
1801 Alexander Bell Drive
Reston, Virginia 20191-4344

Bekey, Ivan.
 Advanced space system concepts and technologies: 2010-2030+ / Ivan
Bekey.
     p. cm.
Includes bibliographical references.
 ISBN 1-884989-12-8
 1. Aerospace engineering--Forecasting. 2. Astronautics--Forecasting.
I. Title.

TL791 .B45 2002
629.4'01'12--dc21

                                        2002153899

Data and information appearing in this book are for informational purposes only.
The publishers and the author are not responsible for any injury or damage resulting from use or reliance, nor do the publishers or the author warrant that use or
reliance will be free from privately owned rights.

The material in this book was reviewed by the U.S. Department of Defense and
approved for public release.

Contents

Preface

During my 48 years as a practicing engineer I have always strived to understand what could be done in the present, given the means and the will. But even more important, I have always looked ahead to exciting future space applications and have then worked hard to advance the technologies required to attain their realization. My fascination has always been with those that others might call "unattainable" or "impractical" to see if I might perhaps be able to find a path to make them real.

I was fortunate to have worked at The Aerospace Corporation during the 1960s and much of the 1970s and then at NASA (National Aeronautics and Space Administration) Headquarters in the 1980s until the late 1990s. My functional responsibilities in various advanced technology and systems as well as my interaction with others working in these areas exposed me to numerous ideas, proposals, and concepts. These in turn influenced many of my own ideas and concepts, which ranged from the unconventional yet near term and modest which required minimal development, to large, ambitious, visionary, and breakthrough concepts representing huge commitments that probably could not be realized for decades.

This book had its effective beginning 25 years ago while I was employed at Aerospace, directing studies and analyses of advanced space systems for military applications in support of the U.S. Air Force. In 1974, NASA engaged Aerospace to perform a far-reaching, long-term study to identify future space applications for both civilian and defense uses and common use high-leverage technologies and space infrastructure that might be supported by both NASA and DOD (Department of Defense). Entitled "Advanced Space System Concepts and Their Orbital Support Needs (1980–2000)," the study completed in 1976 under my direction described high-leverage technologies that could be developed in a 20-year period and a large number of imaginative concepts for space application that could apply these technologies to both future Air Force and NASA missions.

The study findings were enthusiastically received by a great variety of audiences, including the Air Force, other branches of DOD, NASA, and many industry concerns, professional associations, and business groups. Several news media published some of the report's ideas for future space applications. One popular idea was my actual design for a geostationary orbit space system to implement the famous "Dick Tracy" wrist radio, which Motorola eventually incorporated in its Iridium communications satellite concept but using less demanding low orbit satellites and larger hand-held radios. Although the formal purpose of the study was to explore and understand future launch and infrastructure needs that appeared to be common to both NASA and the Air Force, the innovative application concepts and visionary thinking received the most notice and opened new vistas for many people.

Shortly after the study was completed, I joined NASA Headquarters and for the next 19 years held senior management and technical positions planning the future of the U.S. civil space program. For a number of those years, I was director of Advanced Programs in the Office of Space Flight. Upon retiring as director of the Office of Advanced Concepts at the end of 1996, I formed Bekey Designs, Inc., an advanced space systems and technologies consulting firm. Aerospace subsequently retained me to update the 1975 advanced concepts study.

The new study was needed because the prospectus of the 1975 study had reached its horizon—the year 2000—and much progress had been made in space technologies since the study report was first issued. A new look at these technologies would also benefit from my additional experience in civilian and commercial advanced technologies and system concepts gained during my years with NASA.

The new year-long "updated" study looked ahead another 20 to 30 years to the 2020–2030+ time frame to identify high leverage technologies, described the innovative space application concepts they could enable, and discussed the broad implications of their implementation for defense, NASA, and space commerce. The study report was issued July 1998 and its results presented widely, again to enthusiastic audiences.

This book is based on that study, which was a unique experience for me. Having spent more than 40 years advancing the frontiers in both military and civilian advanced space concepts, technologies, and programs, I welcomed the opportunity to "put it all together." It is my hope that this resulting book also represents a useful product—one that offers a visionary yet practical view of what "out of the box" space applications could really become in the next 30+ years, and an identification of some of the new technologies that will enable them, if adequately pursued.

Ivan Bekey

Acknowledgments

The opportunity to create the report on which this book is based came from the vision and support of Dr. George Paulikas, who at the time of the 1998 study was Executive Vice President of The Aerospace Corporation. Most high level executives who are as immersed in the realities of current flight programs and managerial issues as was Dr. Paulikas do not have the vision he had to propose and support a forward looking study of this kind and scope. Also instrumental in reviewing, advising, and supporting the study was Dr. Larry Greenberg, vice president of Aerospace's Laboratory Operations. The opportunity to write this book, based on converting and updating the study report, is owing to the foresight and support of Dr. David Evans, Director of The Aerospace Institute, who envisioned the educational value of the material to a broader audience than The Aerospace Corporation. Thank you, George, Larry, and Dave.

I wish to express my heartfelt thanks to four other persons without which this book could not have become a reality. Grant Aufderhaar, Laurie Henrikson, and Vince Boles, Principal Director, Assistant General Manager, and General Manager of the Advanced Technology Division at The Aerospace Corporation Chantilly Offices who supported and encouraged me to write the book, and to Donna Born, Director of The Aerospace Press of The Aerospace Institute, whose superb editing skills as well as infinite patience and gentleness with my many changes over two years made a quality manuscript possible. Thank you from the bottom of my heart.

Part I

High Leverage Technologies and Techniques

1 Introduction: The Exciting Future for Space

1.1 Linear Thinking Leads to a Gloomy Outlook

The space age—just a little more than 50 years old—has witnessed some astonishing achievements during its relatively short duration. Powerful space communications capabilities, for example, are indispensable to modern life and taken for granted by most users. Indeed, space communications have become ubiquitous technological tools of industrialized society.

Notwithstanding these accomplishments, space applications other than communications have fallen far short of what they could be if adequate attention and resources were invested in new technologies and techniques for exploiting the medium of space. Except for a few notable exceptions such as the Apollo program, development of military, civilian, and commercial capabilities has been a slow, incremental, and conservative evolution with minimal funding and the lowest possible risk. Since the demise of the Apollo program and the Cold War, many bold ideas for utilization of space were advanced but uniformly deferred or rejected. A persistent reluctance to take great steps, push frontiers, and assume risks with untested but high-leverage techniques has essentially prevented many space applications of the Department of Defense (DOD) and the National Aeronautics and Space Administration (NASA) from becoming what they could and should be.

But this scarce implementation of space systems with bold capabilities is not ordained. A number of far-sighted projections of imaginative space applications for both government and commercial use were made even before the 1957 orbiting of Sputnik and were greatly expanded immediately after. A prime example is the 1960s pioneering work by Dr. Krafft Ehricke in space industrialization in which he discussed potential capabilities to provide illumination and power to Earth from space, among other ideas, along with means to attain the requisite launch systems.[1.1] NASA made even more ambitious projections in human spaceflight during and immediately after the Apollo program when it still believed it had a mandate to establish a human presence throughout the solar system. Wernher Von Braun's post-Apollo program included plans for ambitious orbiting outposts as well as bases on the moon and Mars with dozens of astronauts permanently at each location—all to be accomplished before 1981![1.2] Given the large funding sought for these proposals that used tested Apollo technologies, quite likely NASA could have accomplished a sizable portion of those projections—a far cry from the current status.

For its part, the U.S. Air Force over the years has chartered a large number of task forces and special studies to define ambitious future space capabilities, which primarily depended on major advances in a number of technologies, some evolutionary and others revolutionary. Unfortunately, in contrast with the NASA proposals, the technologies for the Air Force and Ehricke studies and proposals were

3

much more difficult and therefore further off than assumed, and in neither case did the hoped-for funding or markets materialize. Thus whether for political, budgetary, or technological reasons, most of the future projections fell far off the mark when they ventured beyond a decade or so into the future, and few have been implemented into actual orbiting space systems.

Every forecast or projection of capability into the future is bound to be erroneous in some important aspects. This is not only because of the limited vision in terms of the possibilities, but because of the utter impossibility of forecasting inventions, especially in the accelerating pace of technology that has characterized the world at least since the industrial revolution. Equally important is the impossibility of correctly gauging the political environment, which can drastically retard or accelerate technological activities. As Edward Teller cautioned more than 30 years ago, "It is extremely hazardous to prognosticate—especially about the future." Think of today's technological innovations that did not exist 50 years ago. Table 1.1 lists technological advancements taken for granted today that either didn't exist 50 years ago or were just in their infancy.

Our world today is vastly different from what most people—except perhaps a few visionaries and many science-fiction writers—would have imagined 50 years ago. Most people (including technical planners) imagine the future to be a more or less "linear extrapolation" from the present. This leads them to conceive future space systems as being similar in form and function as today's systems but having, for example, a greater number of more sensitive detectors, larger antennas of a similar design, larger solar arrays for producing more power, and larger versions of chemically fueled rockets for launch. But just as linear thinking 50 years ago has come up short in predicting technological realities commonplace today because of new inventions and innovations that could not have been forecast, predictions for technological advancements in common use 50 years from now will be even more in error because of the accelerating pace of technology.

For example, consider the pervasive Internet and World Wide Web, where anyone anywhere can immediately access vast information databases and communicate with individuals or groups independently of distance or time zone. Such capability is changing living patterns; commerce at all levels; communications between individuals, companies, and governments; and, in fact, hard-line political regimes. Consider this in the perspective of the average citizen in 1950, who was lucky to have a rotary-dial telephone with a party-line connection, had neither access nor resources to make overseas phone calls, relied on physical libraries for information, used a manual typewriter to type letters in which errors had to be manually erased, and "cut" a stencil to make a few messy mimeograph copies by hand to send out via the U.S. Post Office.

Neither the routine and affordable distribution of hundreds of color television channels via satellite nor the digital Internet/Web technology and its pervasive effects were foreseen, and indeed were all but unimaginable in the context of the average citizen of the 1950s. Nor did many at that time envision the routine airline

Table 1.1. Perspective: Technology 50 Years Ago

Nonexistent	In Their Infancy
Satellites	Radar
Lasers	Transatlantic airlines
Fiber optics	Closed loop control systems
Solid-state electronics	Telemetry
Cell phones	High speed, large memory computers
Jet aircraft	Large screen television
Orbital launch vehicles	Audio tape recorders
Internet/Web	Antibiotics
Desktop computers	Interstate highways
Color television	Copy machines
Organ transplants	Electric watches
Word processors	Walkie-talkies
Microwave ovens	Automatic transmissions
Adaptive optics	Power lawn mowers
Global connectivity/networks	Fluorescent lights
VCRs or FAX machines	Transatlantic/transpacific cables
MRI or CT scanners	Guided missiles

travel by millions of people across the oceans in a few hours in relative comfort for affordable ticket prices, all brought about by development of the jet engine.

Indeed, linear extrapolation with respect to space capability several decades into the future will generate a very gloomy prospect indeed for space and its use:

- The cost of launch into space will not be reduced much from today and will still be several thousand dollars per kilogram of payload. Launch vehicles will still be at least partially expendable.
- Spacecraft weight will not be significantly lower than today for the same function and performance.
- Spacecraft cost will, therefore, continue to be tens of thousands of dollars per kilogram after development investments, often three to ten times as great.
- Power in space will still be costly and limited to a few tens of kilowatts in common use.
- Orbit transfer from low Earth orbit (LEO) to geostationary Earth orbit (GEO) will still be considerably more expensive than lofting into LEO from Earth. The propellant for the transfer will still be brought from Earth and the transferring stage discarded.

- The aperture of practical microwave and optical sensors for Earth-oriented surveillance or astronomical functions will be limited to less than about 30 meters for microwave sensors and less than about 10 meters for optical sensors because of the enormous weight, complexity, difficult packaging for launch and deployment, and cost of the primary apertures as well as the necessary precision structural truss work to maintain their premanufactured accuracy during launch and after deployment.
- The ability to explore the planets and beyond will be limited to small payloads sent infrequently at great expense. Interstellar exploration will be only for the science-fiction writers.
- Military functions of spacecraft will be mainly those that provide force multiplication and are used principally to support the forces applied by ground or airborne means and ground fighting troops.
- Space will continue to be the province of governments, except for the limited commercially successful space communications industry.
- Communications spacecraft will continue to be built for the long life required for unattended economic payoff, and will continue to be expensive and increasingly out of date during their lifetimes because of the inability to access them for servicing or upgrading in orbit.
- Opportunities for humans to travel in space will be limited mainly to government astronauts and highly trained specialists, and a few very rich adventurers.

The aggregate of these characteristics does indeed present a very gloomy prediction. Operating in the space environment would remain an expensive and rare activity engaged in principally by governments, large corporations, and a few rich individuals. In this future world, use of space could never become a part of the everyday life of millions of ordinary citizens as airline travel has become in a few short decades. Much of this gloomy picture is due to the great costs of developing and operating systems for space, which are thousands of times greater than for similar functions performed from aircraft, and because of technological myopia. This book will show—with rational technological innovation and targeted analysis—the fallacy of each of these linear extrapolations and of this aggregate pessimistic view of the future of space.

1.2 New Technologies Will Enable an Exciting Future for Space

This book will describe an infinitely more positive and attractive future for space and the technological wonders it portends. To achieve this promising vision, we must "think out of the box," avoid seductively easy linear extrapolations, and apply inventiveness in those technological areas where breakthroughs will have the greatest effect on our future capabilities and costs. We need not make or postulate any new inventions to bring about the space future we want, although such inventions will surely happen. We must apply every technological area already being worked and identify and apply new technologies that, even though still in the basic research phases, would have great impact if developed and proved.

An imaginative view presented by the book describes what space could well become in the next several decades if new technologies that have already been identified are developed and demonstrated, and bold new innovative applications using these technologies are undertaken. To make that case, the book discusses the future environment for space activities and makes a case for why this environment will be very different from the predominant conditions of the past and present. It identifies a dozen critical technologies with the potential for making orders-of-magnitude reductions in spacecraft and launch vehicle weight and cost, and orders-of-magnitude increases in their performance, if only we would make the investment to develop and prove them. It then uses these technologies to identify and synthesize a large number of space-application concepts, addressing both established as well as unconventional missions and functions that have revolutionary potential.

The emphasis in this book is not on incremental improvement but rather on what has been called "disruptive innovation"—that is, the generation of capability so great, or with cost so low, or both, that revolutionary change occurs. In industry, if ignored, this kind of innovation can be disruptive; if adopted, it can make successful new entities and leapfrog current leaders. While this change is disruptive in that it challenges or threatens current activities and dogma, rapid and great advances in both the private and public sectors generally require such innovation and stem principally from such drastic measures.

1.3 Objectives and Approach

The aim of this book is to show what is possible. The hope is that once people see what is possible, they will raise goals and horizons, challenge political and budgetary limitations, and strive hard to accomplish at least some of the new capabilities—or entirely different innovative capabilities not identified or illustrated here.

Thus emerge the fundamental objectives and approach—the guidelines employed in identifying the technologies and application concepts, which are unchanged from those of the 1975 study of possible future civilian and defense space applications:

- Concentrate on technologies and applications that have the potential for orders-of-magnitude change.
- Avoid 10-percent improvements.
- Aim the horizon far enough into the future to be well beyond the "program Improvement" that is typically the focus of the customers of space systems.
- Pay no attention to current political or policy "correctness" nor to likely required budgets if the capability promises to be unusually useful.
- Collect and innovate concepts that might be useful, but not necessarily those that should be built nor necessarily the most likely to be funded or accepted.

- Do not advocate or get wedded to the concepts—just identify possibilities.
- Innovate, invent, as well as collect application ideas and technologies wherever found.
- Include useful nearer-term ideas and technologies if unconventional, overlooked, or discarded for invalid reasons.
- Remember that "creation of new concepts and capabilities must be its own reward."

1.4 References

1.1. Krafft Ehricke, "Space Industrialization," Rockwell International Space Division, Report No. SD 76-SA-0079-1 (June 29, 1976).

1.2. Wernher Von Braun, "The Post-Apollo Program," Presentation to the NASA Administrator, NASA Archives (1969).

2 General Principles

2.1 Technologies and Techniques for Greatest Leverage on Future Space Systems

Many generic technological trends will enable new capabilities in future systems. Discussed widely in the industry, these include microelectronics, biomolecular systems, nanoscale materials engineering, MEMS (microelectromechanical systems), multifunction techniques, and many others. This book concentrates on a number of specific technologies and techniques that will provide the greatest leverage in future weight, cost, and capabilities of space systems. The space capability applications that will be presented and discussed in Part III are based on these technologies.

Some of these technologies have been known for years, but have not been pursued for a variety of reasons although a number of them will probably come into their own in the near future. Others are fairly new and not widely recognized, but hold revolutionary potential. All of these technologies and techniques are based on solid principles, but the implementation of many is immature and imprecise at this time. The author conceived some, while a number of sources yielded others.

Most of these technologies and techniques will require considerable development and demonstration before they could be seriously considered for adoption into a space system program. Even then, relatively few are likely to be readily accepted or developed by NASA or DOD, partly because they are novel and not well known and partly because the natural tendency of organizations is to develop technologies championed by their own people and already included in their plans. Almost none of the technologies will be developed by industry at its own expense because the payoff is generally too many years into the future.

2.2 High Leverage Principles

Thus, while it is the traditional role of government to support basic research and technology developments of prime relevance to national defense and economic development, such support is not very likely to be readily available unless the payoff of these technologies is well articulated and advocated. A number of these technologies and techniques have exceptionally high leverage in enabling future space system capabilities. These are listed in Table 2.1.

Perhaps the most far-reaching of these principles is the first: "Replace structures with information." It recognizes that relative forces in space are very weak and that the bridge-like trusses and precision shapes universally used in spacecraft today are needed principally while the spacecraft are on the ground and during launch, but not when they reach orbit, where the gravitational forces are essentially zero. Thus heavy, precision trusses and massive apertures designed to hold precise shape on the ground and during launch can be eliminated and replaced with

9

Table 2.1. High-Leverage Principles to Pursue in Space Concepts
"Don't fight the space environment—use it to advantage."

1. Replace structures with information.

2. Adopt distributed space systems. Use coherent cooperation among many spacecraft to implement coherent sparse apertures.

3. Use adaptive gossamer membranes to make large yet lightweight, filled apertures.

4. Fabricate large gossamer membranes in the benign space environment.

5. Transport energy and information, rather than mass, through space.

6. Use spectrally matched multiple bandgap cells and films for high-efficiency solar power.

7. Replace chemical combustion in propulsive devices with electromagnetic and electrostatic forces and plasmas.

8. Exploit electromagnetic, dynamic, and static properties of long tethers.

9. Beam power to remote or difficult to access locations.

10. Service, repair, and upgrade large and complex spacecraft.

11. Leverage the moon's shallow gravity well to mine, manufacture, and transport materials and devices from the moon to Earth orbit and Earth.

12. Exploit the explosion in machine computing, visualization, and artificial intelligence.

13. Utilize designer materials, especially nanomaterials.

14. Exploit nanotechnology, MEMS, and NEMS (nanoelectromechanical systems).

lightweight and thin elements that are shaped only when in space, and spacecraft elements can be flown in a precision formation so that they appear to be held by a virtual structure. This replacement of structural elements by precision formations is made possible by more complex sensing and computation and well controlled but low thrust propulsion.

This principle thus capitalizes on the fundamental attributes of space, where mass is expensive, whereas information and its processing are cheap and lightweight and rapidly getting cheaper and lighter. This principle goes far beyond the so-called controlled structures widely worked in laboratories today, which aim to minimize, but not eliminate, structural mass by intelligent control of reaction masses with actuators.

The ability to perform computationally complex tasks leads directly to the second important principle: "Adopt distributed space systems," in which the whole is more than the sum of the parts. Given that the relative position and velocity of separated spacecraft can be measured and controlled with precision, a distributed group of spacecraft can be made to function as one. In the simplest implementation, many spacecraft cooperate to accomplish a common task. In a somewhat more complex implementation, the relative positions of the spacecraft are precisely controlled so that together they form large coherent antennas or optical apertures without structural trusses. The ultimate implementation coherently combines the signals at or from a large number of separated optical or antenna spacecraft elements whose positions are accurately determined but not accurately controlled. These spacecraft act as the elements of a radio frequency or optical aperture so that the ensemble forms a large coherent array, with dimensions of the total separation. In principle, this will allow implementations of arbitrarily large antennas or optics in space.

The third principle—"Use adaptive gossamer membranes to make large yet lightweight, filled apertures"—capitalizes on the tiny relative forces in orbit to implement large optical or radio frequency reflectors from imprecisely shaped, or initially unshaped, active thin membranes, which are made to assume and maintain the desired shape after reaching orbit. This makes possible very large antennas and optics whose weights are tiny fractions of those, were they to use even the most advanced of today's techniques. In addition, unfolding and shaping a film only after reaching orbit will free the using sensor spacecraft from the constraining limits set by the launch vehicle shroud size because tightly packaged films can deploy to form very large apertures yet controlled to have smooth and accurate surfaces.

The fourth principle goes a step further: "Fabricate large gossamer membranes in the benign space environment." For applications in which a precision film shape is not needed in space, such as for solar sails, the films could be manufactured in space from raw materials brought into orbit. For applications requiring great precision, such as film primary reflectors for antennas or optics, manufacturing the thin films in space could yield far more uniform thicknesses and smoother surface finishes with much simpler manufacturing processes and tools than films manufactured on the ground because gravitational forces are essentially absent in orbit.

The high cost of launching anything into space leads to the fifth principle: "Transport energy and information, rather than mass, through space." Because energy and information weigh nothing, the mass taken to, and transported through, space should be minimized. This principle favors computationally intensive and complex system implementations as they result in lowered total mass, and is a different aspect of the second principle.

The sixth principle, "Use spectrally matched multiple bandgap cells and films for high efficiency solar power," calls for maximizing the efficiency of solar arrays and minimizing their area and weight by using multiple photovoltaic cells and matching their bandgaps to the solar radiation spectrum. In conjunction with the use of membrane concentrators this will reduce weight and cost by minimizing the cell area required for given power levels.

The seventh principle calls for "replacing chemical combustion in propulsive devices with electromagnetic and electrostatic forces, and plasmas." The specific impulse attainable in chemical propulsion is limited by the combustion temperature, while the effective temperature and specific impulse of nonchemical devices can be much greater, decreasing propellant weight while simultaneously improving performance. Whereas such thrusters generally require more electrical energy, the spectrally matched cells and thin films of the above principle will make provision of such energy inexpensive and lightweight.

The eighth principle is to "exploit the electromagnetic, dynamic, and static properties of long tethers." In space, tethers can be extremely long and thin because they are exposed only to relative forces, which in orbit are small. Because they are pure tension structures, the lightest of all, they are extremely lightweight. In addition, tethers are capable of complex dynamics interacting with orbital forces, which can make tethers an elegant means of momentum and energy transfer for changing spacecraft orbits. Furthermore, conductive tethers will interact with the Earth's plasmasphere, generating electromotive forces, making them behave as efficient electric generators or thrust generators in space. These characteristics can further enhance spacecraft flexibility at low weight.

The ninth principle, "Beam power to remote or difficult to access locations," recognizes the unique potential to use space to generate high power and energy beams by conversion of solar energy, and to deliver it to a number of using locations on Earth that are power poor. This avoids the diurnal and weather limits of terrestrial solar power conversion. In addition, such beams could also be generated on Earth and space reflectors used to relay energy to desired receiving locations. The potential also exists for beaming energy from highly efficient centralized power stations in space to a number of spacecraft in space to attain efficiencies of scale in power.

The tenth principle, "Service, repair, and upgrade spacecraft in orbit," avoids the very expensive practice of making throwaway and inaccessible spacecraft systems. As with accessible ground and airborne systems, space systems would be designed to be maintained and upgraded. Lower costs of access to space resulting from application of the other principles will enable economical application of this principle, and in turn result in lower space system costs and far more flexibility.

The eleventh principle is "Leverage the resources and weak gravity of the moon to mine, manufacture, and transport materials and devices from the moon to Earth." Because of the moon's shallow gravity well, less energy is required to launch an item into Earth orbit from the moon than to launch it from the surface of the Earth. Thus, when a lunar infrastructure exists it would be used to supply many needs for materials and devices in Earth orbit.

The twelfth principle, "Exploit the explosion in machine computing, visualization, and artificial intelligence," recognizes that the explosive growth in computing power and the progress in artificial intelligence and related fields for terrestrial applications will provide enough computing support for any space application of these general principles.

The last two principles are also by-products of terrestrially driven applications, which nonetheless will have great effect on space system weight and capabilities. In fact the feasibility of implementing structures and devices grown from carbon nanotubes will have revolutionary potential for space systems because of the extreme importance of low weight in space.

3 Space Apertures at Radio Frequencies and Microwaves

3.1 Radio Frequency Filled Apertures and Antennas in Space

Radio frequency (RF) filled apertures and antennas are currently implemented by a reflecting mesh stretched over a precise deployable truss that determines the desired surface figure of the antenna. While fairly lightweight, this precision structure is based on an approach that demands great design and manufacturing precision, and the storage volume limits the size of the attainable antenna. A new and most powerful technique for space antennas is to shape the reflector not on the ground but only when in space, and then use an adaptive membrane to attain and hold an arbitrary shape in the presence of disturbances. See Fig. 3.1.

The membrane is constructed as a sandwich of two piezoelectric materials, polarized in opposite directions and constituting a bimorph. When a portion of such a membrane, with a set of electrodes deposited on its surfaces, is subjected to a voltage, it responds by bending because one layer gets thinner and larger while the other layer gets thicker and smaller. The result is a localized curvature, which can be used to shape the membrane if the size of the electrodes can be made sufficiently small yet cover the entire membrane. One way of attaining this is to cover both surfaces of the membrane bimorph with a large number of individually addressable pixel electrode elements.

This effect can also be obtained by using an electron beam instead of the front electrodes to define the portion of the membrane to be actuated, and applying a voltage between the back electrode (which extends over the whole surface) and the beam generator.[3.1–3.3] By scanning the beam across the entire surface and modulating the beam current and the back electrode voltage appropriately, a charge distribution

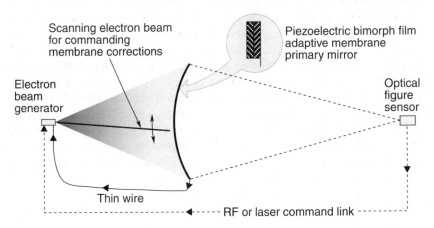

Fig. 3.1. Adaptive membrane shaping and correction. All elements are precisely stationkept with respect to each other in space.

completely equivalent to that deposited by a very large number of pixellated elec-trodes can be deposited within the membrane and can cause it to assume an arbi-trary shape. By making the beam and back voltages responsive to a separated optical figure sensor that determines the actual membrane surface and comparing it to the desired surface figure, correction commands can be sent to the electron beam and back voltage generator that result in a closed loop correction system. This system will take the initially shapeless membrane, cause it to assume a spher-ical, parabolic, or other desired figure, and maintain that figure despite internal or external disturbances. It is a powerful method that will set and control the figure of the primary, both in the gross scale and in removing most small scale figure errors and wrinkles.

The reflector is made from a piezoelectric bimorph film consisting of two layers of piezoelectric film, polarized in opposite directions and bonded together. A third layer consisting of a Nitinol or other shape memory alloy deploys the membrane from an initially folded launch configuration. The membrane could be supported by another film with inflation pressures deploying it, but that is unnecessary and results in an undesirable initial figure that is neither a sphere nor a parabola and at least doubles its weight. The technique described here deploys the membrane totally unstretched and unsupported. It is observed by an optical figure sensor from the front, which determines surface figure errors compared to a reference fig-ure such as a parabola. An electron beam is swept over the back surface and a volt-age is applied to the back of the membrane with respect to the gun via a thin limp wire. The electron charge deposited by the beam and voltage source is modulated in response to commands from the figure sensor error output, and errors are driven nominally to zero via a closed loop process. Elements of this system are station-kept, that is formation flown, to an accuracy less than 1/20 of an RF wavelength.

The use of this technique applied to an RF reflector is illustrated in Fig. 3.2. The technique should permit implementation of filled aperture diameters of at least 100–1000 meters in space. A study performed for NASA by the author[3.4] showed that accuracies to a fraction of a millimeter in the surface should be attainable. The power of the technique is that a precise figure is attained independent of the origi-nal deployed shape of the membrane, and is maintained precisely in the presence of disturbances via the closed loop control system. This is a fundamental and rad-ical improvement over all current and other identified techniques for RF reflectors. The weight of such RF reflectors, calculated utilizing a plated metallic mesh as the reflecting surface and as a function of reasonable film thicknesses, is shown in Fig. 3.3. This weight will be essentially independent of operating frequency because the deposited wire mesh thickness is inversely proportional to frequency, while the mesh spacing is proportional to frequency.

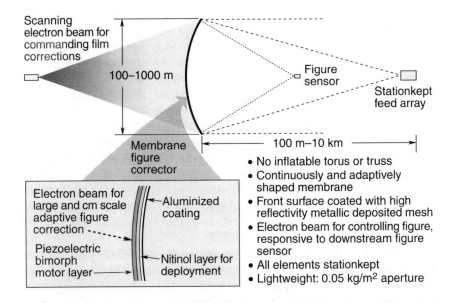

Fig. 3.2. Very large adaptive membrane RF reflector. Surface accuracy is less than 0.5 millimeter root mean square—good to 30 gigahertz.

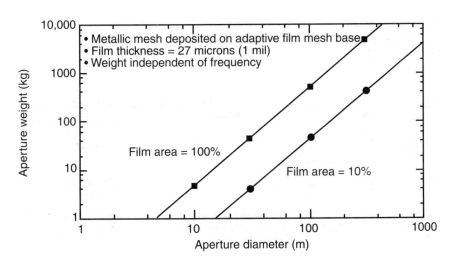

Fig. 3.3. Weight of adaptive RF mesh reflector.

Other new techniques can be employed for implementing large RF antennas as well. A flat membrane can be populated with active elements to produce a full-size phased array. This membrane can be employed as a space-fed array or as a reflecting array ("reflectarray"), operating in conjunction with a feed structure that is either stationkept (formation flown) on or near the axis of the film or is held there by a space tether operating in conjunction with a counterweight on the other side. The membrane can be held flat by spinning it slowly, with or without a complementary spinning mass to cancel the angular momentum, or it could be self-tensioned via the magnetic field self-repulsion caused by a high current flowing in a liquid nitrogen temperature superconducting circumferential wire. In addition, the film could be passive, coated with a Fresnel zone plate, acting in either a transmission or reflection mode. Of course, these films could also be adaptive, as described above, and thus combine active functions with precise closed loop shape control without spinning or superconductive magnetic stiffening. These techniques are illustrated in Fig. 3.4.

The weight of such film-based active arrays will be considerably smaller than that of conventional phased arrays, and will be dominated at higher frequencies by the weight of the active components whose number increases as the square of the frequency. The weight of a membrane-based active reflectarray was calculated using plated wires and elements, foam spacers, and controller chips all directly deposited on the plastic film substrate and ground plane. The total estimated weight is shown in Fig. 3.5.

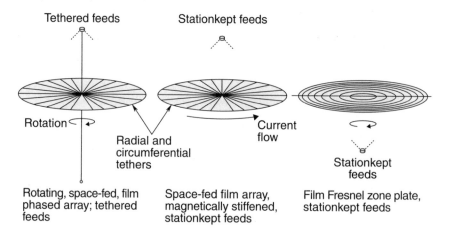

Fig. 3.4. Other large filled RF aperture antennas—potentially 100–1000+ meters in diameter. These could be combined with an adaptive membrane technique.

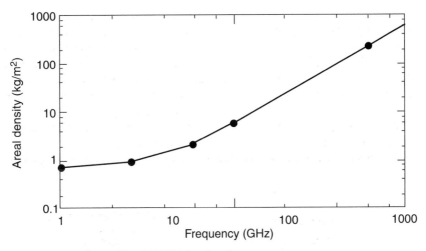

Fig. 3.5. Weight of active space arrays.

3.2 Coherently Cooperating Swarms of Radio Frequency Elements

Swarms of separated elements cooperating coherently can lead to the implementation of extremely large RF antennas for transmission or reception. The elements would be free floating in space, either controlled by propulsive or natural forces for each element to stay within a prescribed "box" in space, or allowed to wander freely so long as they do not stray outside a much larger envelope. The phase or time delay of each element in transmission, reception, or retransmission of a signal would be controlled so that the phase or time delay across the ensemble resulted in coherent addition at a desired point, and the swarm thus acted as a coherent antenna whose dimensions are those of the entire swarm in space.

Because there is no structure connecting the elements, the total weight of a given aperture will be much lower than that of a filled aperture, even if implemented via the adaptive membrane technique discussed in Section 3.1 above. The separation between the elements can be made as small as half of a wavelength if precision controlled, in which case the aperture would function as a filled phased array, indistinguishable in pattern and gain from that of an ideal filled reflector aperture supported by a structure. In principle, however, with complete phasing control of each element, the aperture would be capable of an arbitrary pattern.

The separation between elements can also be made large compared with the wavelength; in which case the swarm will act as a sparse array. In principle, there are no limits to the attainable size of such a sparse array, though in practice some limits will be imposed by orbital dynamics.

The functioning of these coherently cooperating RF elements can be controlled in a number of ways. One is to set up a measuring device such as a lidar to measure the position of each element relative to the feed or focus point to an accuracy of a fraction of the RF wavelength, and command each element in turn to set its phase shift or time delay so the combined wave from all the elements adds in phase. Another is to outfit each element with a receiver of Global Positioning System (GPS)-like signals so that it can determine its own position relative to a reference such as the feed or focus point; each element would control its phase shift or time delay autonomously so as to accomplish the coherent in-phase cooperation of the ensemble without continuous measuring or commanding the elements.

The functioning of these coherently cooperating separated elements can be readily visualized by considering a space-fed phased array, as shown in Fig. 3.6. In a conventional space-fed array the elements contain an input antenna, an amplifier

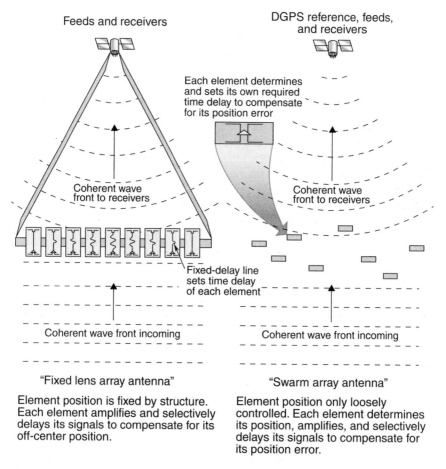

Feeds and receivers

DGPS reference, feeds, and receivers

Each element determines and sets its own required time delay to compensate for its position error

Coherent wave front to receivers

Coherent wave front to receivers

Fixed-delay line sets time delay of each element

Coherent wave front incoming

Coherent wave front incoming

"Fixed lens array antenna"

Element position is fixed by structure. Each element amplifies and selectively delays its signals to compensate for its off-center position.

"Swarm array antenna"

Element position only loosely controlled. Each element determines its position, amplifies, and selectively delays its signals to compensate for its position error.

Fig. 3.6. Coherent swarms of RF elements. Space-fed receiving lens antenna is shown as an example.

and/or phase shifter, and an output antenna. The elements are held rigidly along a plane. Each element has a phased shifter or delay line, and the fixed delays are adjusted so as to convert a spherical wave to a planar wave, or vice versa. The feed is held by structure fixed relative to the elements.

The swarm antenna dispenses with the planar structure as well as the truss structure, and each element is free to wander inside a "box" in space whose dimensions are small compared to the overall antenna diameter but generally larger than the wavelength. Each element is commanded (or is self-directed) to apply a phase or time delay correction that will just compensate for the difference in position between the element's desired and actual positions. This control then corrects the phase so that all elements add coherently, regardless of their actual positions. The swarm then acts exactly as does a conventional structure, fixed lens antenna.

If the elements are positioned in space by propulsive means, then in principle they could be positioned precisely along a plane or sphere, minimizing or eliminating any need for them to have phase or time delay correction means. Whether to employ position control, phase/time control, or some combination of both is dependent on the application and implementation details.

The limits of error in position of the elements and size of the antenna obtainable are very much dependent on a number of considerations, and will be much larger in a direction parallel to the antenna axis than perpendicular to it. Nonetheless the principle is fundamental. In Part III, the principle will be applied to implement a number of different concepts that would be impossible or impractical with structure-based arrays.

The orbital forces acting on the separated elements of the swarm and the feed will in general cause the constellation to disperse rapidly if not constrained because each element is in a slightly different Keplerian orbit, and the orbital element differences will cause dispersion in time. Their positions, whether precise or approximate, can be held by propulsive means, but this is usually practical only for small separations, short times, or both, as the total impulse required to fly non-Keplerian orbits can become large. The solution is to develop very high specific impulse micropropulsion for such stationkeeping, to use tethers to hold the relative positions without propulsion, or some combination of the two. Both techniques will be discussed further in this chapter.

There exists a particularly appealing technique for avoiding much of the impulse requirements for maintaining a constellation by propulsive means in orbit that capitalizes on the mechanics of relative orbits. This technique, though not new having been used for the first GPS concepts in the early 1960s as well as by NASA for an early "coorbiting companion" for the International Space Station, is not well understood by many. It capitalizes on the mechanics of a body in orbit being given a slight displacement velocity relative to an unperturbed body in the same orbit but displaced in track. Though the perturbed body will go into a different orbit with different orbital parameters, its trajectory relative to the unperturbed body will be an ellipse with the unperturbed body at its center. Thus in relative coordinates the perturbed body flies a trajectory around the unperturbed one. This is not merely an illusion but a factual and observable trajectory. If many elements are given slight ΔV impulses (changes in velocity) at different initial separations, they will lie on the apparent surface of a plane rotating about a center in the original orbit. See Fig. 3.7.

This extremely powerful concept, elaborated by Siegfried Janson of The Aerospace Corporation, results in a planar distribution of elements that appear to rotate as a solid disk about its center, with the elements maintaining their relative positions without stationkeeping propulsion. In time, the differential forces and third-body effects will result in relative drifts, requiring propulsion to maintain station for each element, but these ΔVs are an order of magnitude smaller than would be required to maintain the same constellation with propulsion alone. This is an ideal example of working with the space environment, rather than fighting it.

This technique will result in sparse array antennas of perhaps tens, if not hundreds, of kilometers across, implemented with no structure and yet functioning as a coherent array. They will be orders of magnitude lighter than if implemented via structural techniques. Applications of such sparse array antennas will be discussed extensively in Part III.

Thus the ability to coherently combine the signals to or from a swarm of elements can take several forms, as illustrated in Fig. 3.8. Some of these forms include attaching the elements to a membrane with the feed structure tethered through its center, free-flying the elements with the feed structure also free-flying on the axis, placing the elements into a rotating virtual disk with the feed structure stationkept or tethered along the axis, or some combination of these techniques to suit a desired application. The individual elements can be dipoles, subarrays, or individual subapertures, either reflectors or phased arrays.

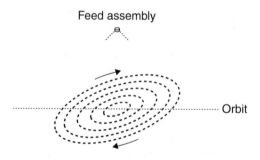

Feed assembly

Orbit

• Elements in stable, planar, circulating relative orbits
• Only second order ΔV required to maintain the constellation

Fig. 3.7. Large sparse "circulating orbit" RF space arrays with no structure. Element subarrays equal 0.1–10 meters; ensemble diameter equals 10–100 kilometers.

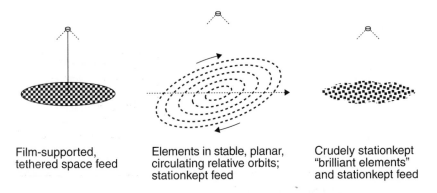

Film-supported,
tethered space feed

Elements in stable, planar,
circulating relative orbits;
stationkept feed

Crudely stationkept
"brilliant elements"
and stationkept feed

Fig. 3.8. Large sparse RF space-fed arrays with little or no structure. The feed could be stationkept or tethered in each use. Element apertures equal 1/2–1000 meters; ensemble diameter equals 10–100 kilometers.

The resultant space array diameter can be extremely large, limited only by non-linear effects due to orbit dynamics. The angle the rotating disk takes with respect to the orbital plane can be adjusted by proper choice of ΔV parameters, so that the disk has a finite diameter at the aspect angle desired for any particular application, from being in the vertical to the horizontal plane. The stability of the element position with time is maximized for a particular angle, and is proprietary to The Aerospace Corporation.

In view of the foregoing treatment of advanced antenna techniques, a number of generic implementations can be classified according to types, as is illustrated in Fig. 3.9. These are not mutually exclusive and can be applied in combinations, but they do serve to illustrate that a number of potential paths exist for evolution from today's structure-defined antennas. Each of these directions affords a degree of freedom from size and weight constraints of current practice that cannot be achieved by linear improvements to current techniques, but rather requires completely different paradigms such as the replacement of structures by information and control. This perfectly illustrates some of the high leverage techniques that will be used in the applications to follow.

These techniques can carry to its ultimate the coherent cooperation of many unconnected elements by the following extrapolation. If a swarm of elements can be caused to coherently cooperate and form an aperture perhaps thousands of kilometers across, then in principle, different swarms located in different orbits can also be caused to coherently cooperate and form antenna elements separated by millions of kilometers. Of course the task of cohering them grows at least linearly with the separation distance, requiring the phase shifts or time delays attained to grow by the same factor, which may be very difficult to attain in practice. Nonetheless in principle this can be done so that separate swarms coherently cooperate

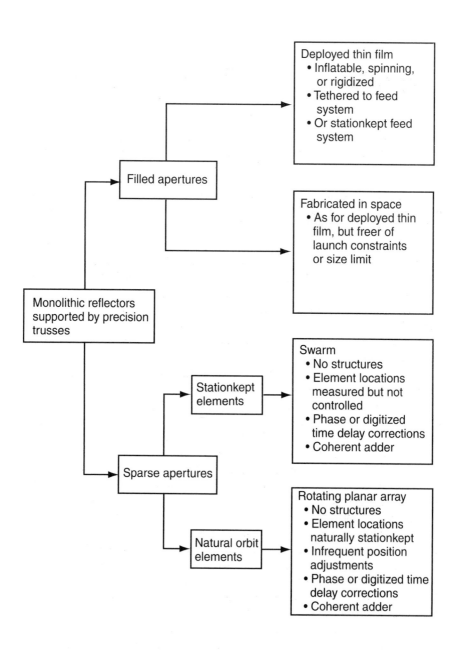

Fig. 3.9. Large RF apertures in space.

to form a much larger effective unit, with coherence not only within a swarm but between swarms as well. This could allow adding power from all elements on a single spot on the ground, for example, or increasing the sensitivity and selectivity of a receiving aperture. These possible swarms are illustrated in Fig. 3.10.

The ultimate achievement in this trend of cooperating RF systems is shown by the right diagram in Fig. 3.10. It is a truly distributed space system in which a myriad of tiny and inexpensive, identical, RF elements or units are placed into a rough orbital belt. Groups of elements within this "orbital swarm" can be programmed or commanded to function coherently, and different functions created in different physical regions. Thus, for example, a communications function can be created over one geographical region while a sensing function can be implemented over another. However, whereas functions so created are geographically fixed, the elements that create them are constantly changing as the satellites move in their orbits through the different geographical areas. The identical elements thus perform different functions depending on their instantaneous locations.

Because each spacecraft is identical even though it performs cooperative everchanging functions commanded or set by software, this swarm is a truly "distributed space system." This contrasts with what is euphemistically called distributed

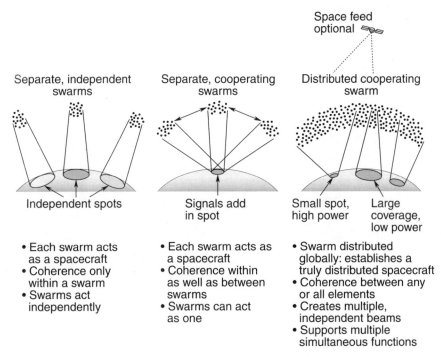

Fig. 3.10. Distributed RF swarms. All swarms are composed of "brilliant elements" with each element independently controlled.

space systems today, which are not much more than proliferated systems in which a large satellite's function is replicated by many smaller ones with little cooperation or coherent addition among them. In contrast, because each space-craft in an orbital swarm is identical yet can perform any of a set of desired functions that can be implemented whenever and wherever summoned by the software, this configuration is perhaps the ultimate achievement in the goal of "anti-stove-piping" space systems.

3.3 References

3.1. NASA Advanced Concepts Research Program, Contract No. NAG8-1317 to John Main. University of Maine.

3.2. John Main, "Smart Material Control Without Wire Leads or Electrodes: New Methods and Devices," *Conference Proceedings, ASME Design Engineering Technical Conference* (Sacramento, CA, September 14–17, 1997), ASME Paper No. DETC 97/VIB-3933.

3.3. Tammy D. Hansen, et al., "Thin Skin Deployable Mirrors for Remote Sensing Systems," Sandia National Laboratories Report No. SAND2001-0101 (January 2001).

3.4. "An Extremely Large Yet Ultra Lightweight Space Telescope and Array: Feasibility Assessment of a New Concept," Grant # 07600-006, NASA Institute of Advanced Concepts (May 29, 1999).

4 Space Optical Apertures and Systems

4.1 Filled Aperture Optical Systems in Space

The basic imaging system concept introduced here embodies the powerful new principle for space systems: "Replace structures with information." The first principle in Table 2.1, it capitalizes on the fundamental attributes of space, in which mass is costly, but information and its processing are lightweight and cheap, and rapidly getting cheaper.

The imaging system concept describes an initially shapeless membrane primary mirror that is given a precision shape only when in space, and a virtual truss rather than precision structure for precisely positioning the elements of the system relative to each other. The lack of a precision-ground, solid, heavy glass or composite primary mirror whose figure must be precisely ground and maintained through launch and the absence of a large and heavy precision truss structure make for an extremely lightweight system. Information-intensive methods to adaptively shape the membrane and to stationkeep all the elements with a virtual truss are substituted for the solid primary and the truss structure.

Space telescopes have been evolving in this direction, but none have yet made the final leap, as does this concept. This evolution is illustrated in Fig. 4.1. The 2.5-meter Hubble Space Telescope uses a large precision truss and thick glass primary, and while it has attained phenomenal performance, it weighs about 12,000 kilograms, which means its total areal density is about 2500 kilograms per square meter of aperture. The new 8-meter NASA James Webb Space Telescope (JWST) will use lighter weight composite, deployable mirror segments and a lightweight composite deployable truss, and its areal density will be two orders of magnitude less: 25–50 kilograms per square meter of aperture. A number of inflatable

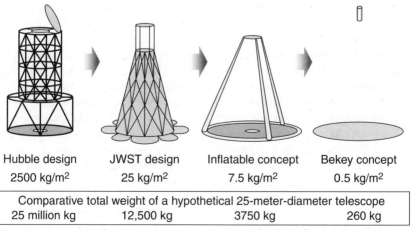

Hubble design	JWST design	Inflatable concept	Bekey concept
2500 kg/m²	25 kg/m²	7.5 kg/m²	0.5 kg/m²

Comparative total weight of a hypothetical 25-meter-diameter telescope			
25 million kg	12,500 kg	3750 kg	260 kg

Fig. 4.1. Evolution of optical space telescopes.

membrane mirror concepts with inflatable, controlled trusses have been studied, but none are defined to the point of demonstrating even conceptual feasibility; however, if they could be made to work, their areal density would probably be at best in the order of 6–10 kilograms per square meter of aperture.

While the weight of the truss has been decreasing with these evolutions, none of these designs embrace the inevitable: that there is no need for any truss in space. Telescope trusses are a carryover of Earth-bound thinking. Since only very weak relative gravitational forces have to be resisted in space, a truss is not needed to hold the elements at precise separations, and precision stationkeeping can be used just as well to form a virtual truss. In addition, in the large primary membrane mirror, we can also substitute information for mass if a thin film mirror is used and its shape is not precision predetermined on the ground but rather commanded by software in space. Both the stationkeeping and the shape adjustment can then be made responsive to outside disturbances, and closed loop correction control can be introduced. Thus the imaging telescope's separated parts and flimsy membrane can be maintained in a virtual structure, which, if sufficiently accurate, is indistinguishable from that of a heavy, fixed precision truss and solid mirror.

Because, without heroic measures, this technique is unlikely to remove all membrane surface errors to attain an optical quality surface, a second stage of correction consisting of a liquid crystal spatial light modulator driven by the remaining errors in the primary corrects the passing light to attain optically coherent wave fronts. In order to function, this liquid crystal must be located at a place in the optical train where a real image of the surface of the primary mirror exists.

Feasibility calculations[4.1] performed by the author as president of Bekey Designs, Inc., in 2000–2001 indicated that the total on-orbit weight of this new concept should be about 260 kilograms for a 25-meter-diameter clear aperture, thus weighing about 0.5 kilograms/meter^2 for the entire telescope (not just for the primary mirror, which weighs 0.07 kilograms/meter^2). In addition, the absence of a structural truss makes the system's weight essentially independent of distance between the elements, so that large focal length systems are as easily attained as short ones without a weight penalty.

To attain this promise, the imaging system employs a number of critical features, listed in Table 4.1. A piezoelectric bimorph membrane primary mirror fully adaptive across its surface is initially shapeless and limp, similar to a plastic-wrap-like film. It is deliberately not inflated, but free floating with no tension applied. As it is adaptive throughout, its surface is shaped into the precise figure in space under the action of an electron beam scanned across its back surface and a voltage applied between the beam and the membrane back surface electrode. This action is analogous to that described for RF apertures and was illustrated in Fig. 3.2. The actual figure of the mirror membrane is measured by a stationkept figure sensor, and the measured errors with respect to a reference figure are converted into shaping commands, which are sent to the electron beam generator. The figure sensor, membrane, and electron beam generator form a closed loop, making the attained

Table 4.1. Concept Principles
"Replace Structures with Information"

====

- No truss structure—precision stationkeep all elements (formation flying)
- Initially shapeless primary mirror
  - Limp plastic-wrap-like piezoelectric bimorph membrane
  - Uninflated, unsupported, free
  - Adaptive throughout its surface
  - Shaped into a precise figure by electron beam only when in space
  - Shape maintained by closed-loop figure control system
- Liquid crystal second-stage corrector to take out remaining errors
- An extremely lightweight, inexpensive, easy-to-build system

====

figure resistant to disturbances and correcting initial wrinkles or other surface imperfections. The telescope implementation is illustrated in Fig. 4.2. Shown in the figure are the two stages of correction, the stationkept (formation-flown) elements, and the lack of a truss structure. The illustration is schematic and not to scale, but assumes that an f/10 system is implemented—that is, the focal length is 10 times the diameter of the primary.

A point design undertaken in the feasibility study included a further innovation: using a tether for rough positioning of the elements of an actual imaging system in geosynchronous Earth orbit (GEO), with propulsive fine stationkeeping. The tether avoids the propellant that would otherwise be needed for the elements to fly non-Keplerian orbits continually. In addition, a number of other innovations were implemented, such as fast field of view (FOV) steering using stationkept small mirrors near the focal assembly. These and other design innovations are contained in the feasibility calculations. The tethered configuration of that study is shown in Fig. 4.3, which illustrates how the basic concept differs from conventional space telescopes and imaging sensors and why the telescope spacecraft in the concept is so lightweight.

The weight of filled aperture space telescopes incorporating the adaptive membrane and stationkept elements technique was calculated, and a point design for a 25-meter-diameter clear aperture imaging telescope in GEO showed the total weight to be only 260 kilograms, which demonstrates the revolutionary potential of this concept. Fig. 4.4 shows the weights scaled for other telescope or imaging sensor diameters. The filled aperture weights are more than an order of magnitude lower than those attained using the best contemporary design technique—that of the NASA James Webb Space Telescope (JWST) at an 8-meter diameter—and would retain that advantage even at 100–300-meter diameters. A 100-meter-diameter, filled aperture telescope using the adaptive membrane and stationkept elements design is estimated to weigh only 1800 kilograms in space,

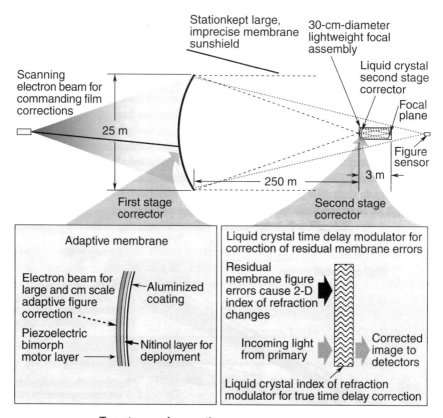

- Two stages of correction
  - adaptive membrane
  - 1000-wave liquid crystal
- No truss structure
- Elements stationkept to ≈ 0.1 mm
- Large focal length attainable with no weight penality
- Imaging at visible light or infrared

Fig. 4.2. Two-stage corrected, adaptive, structureless telescope.

compared with more than 200,000 kilograms estimated using a JWST-type design. This is indeed revolutionary.

A different point design estimated a total weight of 1600 kilograms for a 250-meter-diameter sparse imager that had an annular continuously adaptive surface primary developed using the same design principles but with only 10 percent of the primary area populated by the reflector. That weight is also shown in Fig. 4.4. The sparse-aperture weights are yet another half an order of magnitude lower than those of the adaptive membrane designs using filled apertures. This is a perfect example of the power of the high leverage principles noted in Chapter 2.

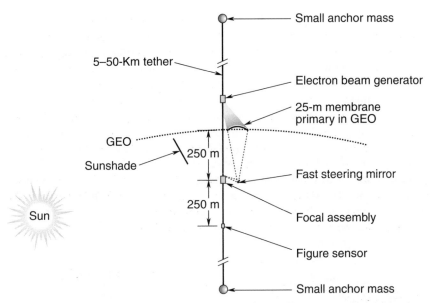

Fig. 4.3. GEO-deployed configuration using a tether for coarse positioning of the elements.

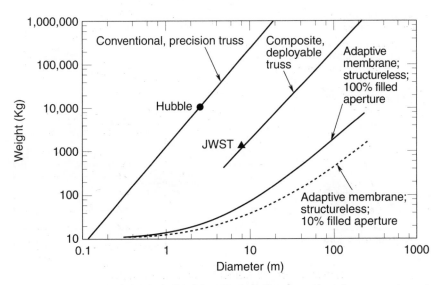

Fig. 4.4. Weight of space optical telescope; complete imaging telescope systems in orbit.

4.2 Swarms of Cooperating Optical Elements

The adaptive membrane primary and stationkept elements technique is fully applicable to sparse aperture systems. A number of adaptive membrane primary apertures can be stationkept in space along the surface of a virtual paraboloid. Each aperture can be shaped to be a segment of a larger parabola (or sphere or other figure) on command. Images of these primaries can each be independently corrected at appropriate locations in a focal assembly, and then the separate light paths added in phase at one or separate focal-plane arrays. In this way, a sparse aperture imager is created whose diameter can be much larger than that of any individual aperture. Because the accuracy of positional control of the separated apertures should be essentially independent of their separation distance, apertures 300 meters to 300 kilometers should be feasible. Spacing of the apertures can be varied in use, with their shapes modulated as needed to maintain the appropriate figure for their location. In this way zooming space telescopes can be readily implemented. See Fig. 4.5.

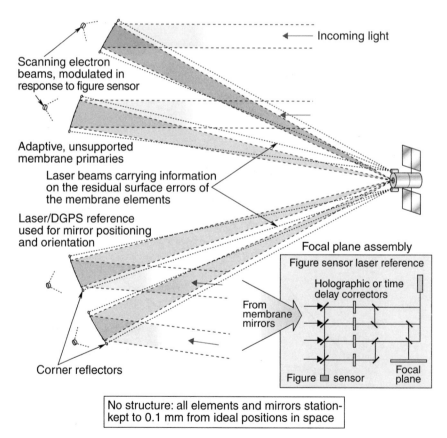

Fig. 4.5. Coherent swarms of optical elements; sparse aperture 300-meter to 300-kilometer imaging optical sensor application.

These sparse aperture optical systems will be much lighter and cheaper than conventional or even inflatable concepts. They should therefore make practical the implementation of NASA's Origins Program, which calls for the ultimate fielding of 25 telescopes, each 25 meters in diameter, in a ring hundreds of kilometers in diameter in order to function as an interferometer to image Earth-sized planets around nearby stars.

The ability to produce primary reflectors (or complete telescopes) in diameters of tens to hundreds of meters also allows these apertures to be localized not only by propulsive stationkeeping, but also by techniques completely analogous to those described for RF apertures in Chapter 3. Thus the apertures could be attached to a large but inaccurate membrane or caused to revolve around a central point without primary stationkeeping propulsion, and attain very large diameter sparse optical arrays. The focal assemblies could be propulsively stationkept, or tethered along the optical axis, or some combination of both, as illustrated in Fig. 4.6.

The adaptive membrane technique allows large apertures to be folded for compact launch, with most of the wrinkles removed after self-deployment. The performance of the systems will still require a high degree of surface smoothness to avoid excessive scatter. However, in principle, by taking advantage of the near zero gravity forces in orbit, it should be possible to fabricate in space the needed membranes in essentially perfect spheres with extremely smooth surfaces. One method, illustrated in Fig. 4.7, inflates a large spherical balloon from raw materials in space; the balloon will be extremely smooth and accurate because of the absence of essentially all convection and gravitational forces. A smaller inflated torus could be bonded to the larger balloon and the undesired portion cut off, leaving a spherical membrane of arbitrary size. Vapor deposition would then deposit reflecting coatings, and other operations would be performed to obtain a perfect

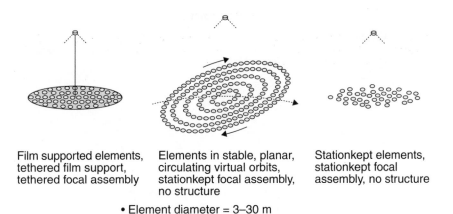

Film supported elements, tethered film support, tethered focal assembly

Elements in stable, planar, circulating virtual orbits, stationkept focal assembly, no structure

Stationkept elements, stationkept focal assembly, no structure

- Element diameter = 3–30 m
- Ensemble diameter = 300 m to 300 km

Fig. 4.6. Large sparse optical arrays with little or no structure.

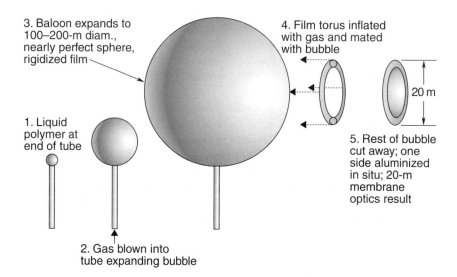

3. Baloon expands to 100–200-m diam., nearly perfect sphere, rigidized film

4. Film torus inflated with gas and mated with bubble

20 m

1. Liquid polymer at end of tube

5. Rest of bubble cut away; one side aluminized in situ; 20-m membrane optics result

2. Gas blown into tube expanding bubble

Fig. 4.7. Fabricating 20-meter membrane reflector in space avoids surface irregularities otherwise inevitable (due to rolling, folding, and deployment).

spherical reflector completely smooth with no deployment wrinkles and a diameter much larger than that of any launch vehicle.

The availability of some or all of these technologies and techniques will make feasible system concepts employing large yet lightweight, and therefore probably very cheap, optical instruments. A large number of imaginative applications of such telescopes and imaging sensors is presented in Part III.

4.3 References

[4.1] Ivan Bekey, Bekey Designs, Inc., "Very Large Yet Extremely Lightweight Space Imaging Systems," *SPIE Astronomical Telescopes and Instrumentation Conference* (Waikolo, Hawaii, Aug. 22–28, 2002), SPIE Paper No. 3543.

5 Tethers in Space

5.1 Tethers

Tethers are the lightest of all structures because they operate in pure tension, and all structural materials are the strongest in tension. Tethers are simple yet elegant and lend themselves to a great many uses in space. Some have already been mentioned in the preceding chapters on radio frequency and optical arrays.

Tethers have flown in space 17 times as of the date of this report, and why they have not yet been applied seriously in space systems is a mystery. Perhaps they are too simple, and engineers prefer complex problems to solve; however, though simple in principle, tethers do have complex dynamics. With understanding and proper design, deleterious effects of these dynamics can usually be avoided, much like structural modes in solid structures. Thus, engineers should get excited about solving tether dynamics problems, so there is hope yet.

Because they are tension-only structures and because gravity gradient forces in near Earth orbits are small, tethers can be long yet lightweight. To understand this phenomenon, imagine a tether suspending two masses at its ends aligned with the local gravity gradient. In low orbit and for contemporary Spectra or Kevlar materials, the length at which the tether's weight equals the weight of the payload masses at its ends—"the critical length"—is about 300 kilometers. In geostationary orbit, because of the much weaker gravitational field, the length increases to about 5000 kilometers.

5.2 Tether Applications

A number of applications of tethers are illustrated in Fig. 5.1. Tethers can hold two masses (payloads or one payload and a countermass) aligned with the local vertical. In the process of deploying from an initially adjacent configuration, energy and momentum are transferred from the lower mass to the upper mass, with their sum remaining constant. If the tether is then cut or released, the upper mass will be in an orbit with higher apogee, having excess energy and momentum for its orbital altitude at release. The lower mass will be in an orbit with lower perigee, having a deficit of energy and momentum for its orbital altitude at release.

In this way payloads can be maneuvered in space without expending energy or propellants. The attainable difference in altitude between the lower and upper masses half an orbit after release is seven times the tether length at release, if released from a stable local vertical orientation. If released while librating over 180 degrees, the difference in altitudes is 14 times the tether length. This means that in low Earth orbit (LEO), altitude differences of up to 4000 kilometers are attainable, while in GEO this difference rises to 70,000 kilometers—more than the GEO altitude itself. These differences are smaller for tether lengths approaching the critical length, since the tether weight then approaches that of the end masses, yet nevertheless can result in highly useful capabilities.

If the masses are set to deliberately spin at more than the orbital rate and then released, they can be imparted a velocity increase (ΔV) of up to about 1000 meters per

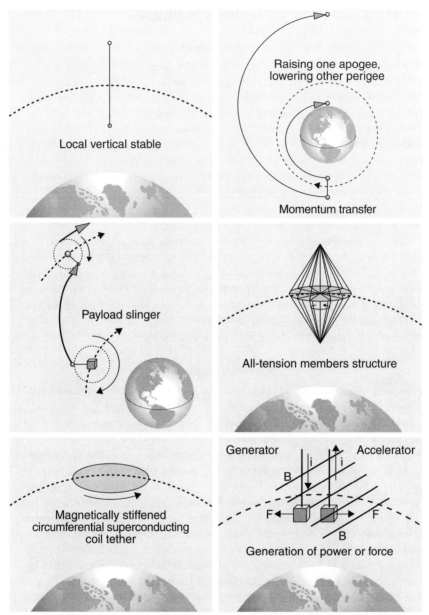

Fig. 5.1. Applications of tethers in space: i = current; B = magnetic field; F = force.

second each with strength/weight ratios of contemporary materials. A "tethered launching platform" in LEO can sling payloads to apogees at much greater altitudes, where they can be caught by a "high altitude tethered catching platform," with proper timing and both platforms spinning. Because the same velocity will be gained by the payload in a release operation as in the catching operation by the upper tether, payloads can be caused to gain three to four times the maximum set by the tether materials for a single operation, and can achieve a total ΔV of 3000–4000 meters/second with current materials. By exploiting this dynamic momentum exchange, payloads can be thus maneuvered between most Earth orbits without requiring any propellants. A specific study[5.1] has determined that maneuvering a payload between eccentric LEO and GEO is possible with one stage, using existing materials. As material strength/weight ratios improve, circular LEO to GEO transfer with one stage should be readily attained with low weight tethers, resulting in inexpensive, reusable, propellantless transfer to and from GEO. This propellantless transfer is illustrated in Fig. 5.2. When tethers are made of conducting

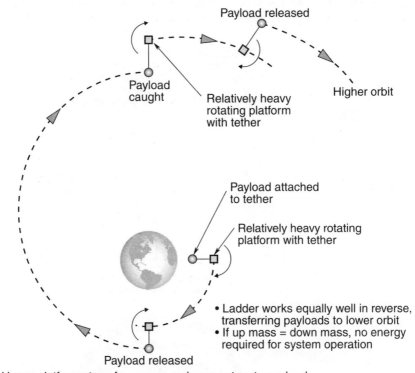

- Heavy platforms transfer energy and momentum to payload
- Each platform adds velocity to payload
  - Tangential velocity subtracts from orbital when catching and adds when releasing payload
  - "Catches" are at zero relative velocity
- Payload placed into Hohmann transfer trajectory
- No propellants expended in transfer
- Platforms make up lost energy with ion propulsion over longer time
- Effective specific impulse of transfer is 5000 seconds, yet transfer times are like those using chemical propulsion (6 hours from LEO to GEO)

Fig. 5.2. Fast orbit transfer without propellants.

wires, insulated for most of their length, they can generate electrical power by inter-acting with the Earth's magnetic field at the expense of orbital energy, or they can generate electromagnetic thrust at the expense of solar array power. Thus they are reversible motor-generators. By timing the current flow, any one or all of the orbital elements of a spacecraft can be changed using such electromagnetic tethers.

Large spacecraft constellations can also be constructed using tethers in pure ten-sion if slow rotation about one axis is adopted. In addition to these effects, large constellations can be formed by exploiting the divergence of the gravity field with altitude. Constellations tens of thousands of kilometers across can be imple-mented, as shown in Fig. 5.3. In this application all elements are in tension except the lower strut, which must be a compression member.

The function of a compression member could be handled using a novel tech-nique, illustrated in Fig. 5.4, or a large conventional space truss, if its weight were

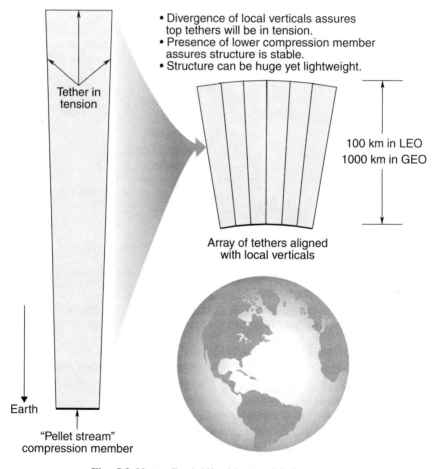

Fig. 5.3. Naturally rigidized large orbital structure

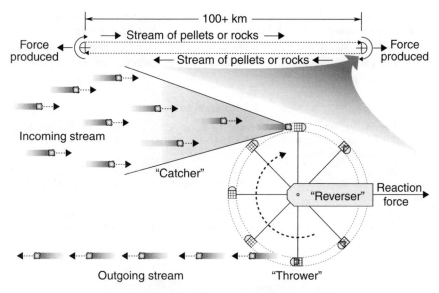

Fig. 5.4. Ultralight, very long compression "truss."

acceptable. The novel technique that resists compression forces without any structure consists of basically throwing rocks or other masses back and forth between the ends of the device. Reaction forces to the act of throwing and catching the masses create a separating force, which is indistinguishable from that exhibited by a solid strut, and thus is able to withstand compression forces.

Many uses of the principles of space tethers have been identified. A few such applications for illustrative purposes are presented in Table 5.1. Recently a number of applications were studied[5.1] for "throwing" payloads between LEO and GEO, between LEO and the moon, and more recently[5.2] between LEO and Mars. Other studies have looked at demanding applications in which the unique capabilities

Table 5.1. A Sampler of Uses of Tether Principles

Application	Near Term	Midterm	Far Term
Constellations	Very large sparse arrays	Variable gravity facilities	Earth–GEO elevator
Momentum transfer	Launcher payload increase	Orbit transfers	LEO–GEO, moon, Mars transfers
Electrodynamic	Stationkeeping and drag makeup	Maneuvering states and power sources	Electrodynamic decelerators
Aerodynamic	Mach-25 long-duration wind tunnel	Mars atmospheric sample return	Tether-enhanced aerodynamic turns

of space tethers are used to advantage in reducing transfer time, increasing payload, or both, compared with propulsive transfers. A NASA tether applications handbook[5.3] contains these and many other applications of tethers in space. A number of imaginative applications of space tethers, both for defense and civilian space use, are presented in Part III.

5.3 References

[5.1.] Robert Hoyt (Tethers Unlimited, Inc.), "Cislunar Tether Transportation System," Final Report, NIAC Contract 07600-011, NASA Institute of Advanced Concepts (May 17, 1999), http://niac.usra.edu/studies.

[5.2.] Robert Hoyt (Tethers Unlimited, Inc.),"Moon and Mars Orbiting Spinning Tether Transport," Final Report, NASA IAC contract #07600-034 (September 18, 2001), http://niac.usra.edu/studies.

[5.3.] M. S. Cosmo and E. C. Loprenzine, editors, *NASA Tethers in Space Handbook*, 3rd edition (NASA Marshall Space Flight Center, Huntsville, Alabama, December 1977).

6 Solar Arrays and Thermal Radiators

Current single bandgap solar arrays are very inefficient in converting solar radiation into electrical power. Much of the incident energy is at wavelengths longer than the cell bandgap energy and produces no current, or the wavelengths are much shorter and a portion of the incident energy is wasted because it does not produce extra current. Vertically stacked cell arrays of two and even three bandgap energy cells have been produced to more fully utilize the incident solar spectrum. Arrays that stack three to four types have shown efficiencies of 28–35 percent, but the voltages of the different cells are difficult to match because all cell types must be the same physical size yet produce different voltages. In addition, the cells cannot be used with very high concentration ratios due to heat dissipation limits because the area for heat rejection is the cell area and is unchanged from single bandgap cell arrays.

A technique conceived by Dr. Wade Blocker of The Aerospace Corporation in 1976, and investigated under the direction of the author and Dr. Neville Marzwell of NASA's Jet Propulsion Laboratory (JPL), promises to be much more efficient and lighter. This technique splits the incoming solar spectrum into a continuum rainbow and impresses it on a number of separate cells, each of whose bandgap is tailored to best match the incoming narrow range of wavelengths impressed on it. (See Fig. 6.1.) Work done at JPL indicates that new cell materials do not have to be designed, as combinations of existing cell materials will result in good spectral matching characteristics. The cell types required exist: principally gallium indium phosphide and aluminum gallium arsenide in various dopings, silicon carbide, silicon, and germanium. These are real cells, though some types have not yet been production-engineered for large quantities. In addition, a much greater number of spectral regions can be used than is practical with vertically stacked cells, increasing the efficiency of conversion. Table 6.1 lists cell materials for three to nine cell systems.

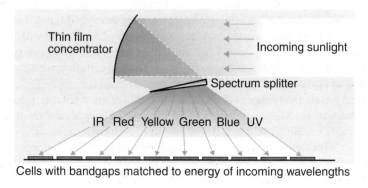

Fig. 6.1. Rainbow array concept.

Table 6.1. Cell Materials
(Each cell type is chosen to optimize conversion of the vacuum solar spectrum.)

Three-Cell System		Six-Cell System		Nine-Cell System	
Material	Electron Volts	Material	Electron Volts	Material	Electron Volts
AlGaAs	1.94	GaP	2.26	6H-SiC	2.90
GaAs	1.43	AlGaAs	1.94	3C-SiC	2.30
InGaAs	0.74	InGaP$_2$	1.77	Ga$_{.53}$In$_{.47}$P	1.95
		GaAs	1.43	Al$_{.2}$Ga$_{.8}$As	1.71
		InGaAs	0.74	GaAS	1.43
		InGaAs	0.62	Ga$_{.87}$In$_{.13}$As	1.25
				Si	1.11
				Ga$_{.48}$In$_{.52}$As	0.75
				Ga$_{.36}$In$_{.64}$As	0.60

Because each cell type's size and shape can differ and a different number of such cells can be connected in series, a good voltage match can be attained for the different cell types, resulting in good power matching and fewer losses than possible with vertically stacked cells. This results in higher efficiency for the same number of spectral regions. (See Fig. 6.2.) Additionally, this technique minimizes the heat load on each cell because most of the spectrum is diverted elsewhere, and each cell has its own area for thermal dissipation. Thus higher concentration ratios can be used than for vertically stacked cells and the cell temperatures kept low, which also maximizes the efficiency.

The net effect of all these features is that the resulting "rainbow" solar array has a theoretical efficiency approaching 70 percent using many cell types. Laboratory experiments by JPL have already yielded an efficiency of 52 percent with four cell types and are expected to reach 65 percent with a larger number of cell types and concentration.[6.1] This is in contrast to vertically stacked cells that are not expected to exceed an efficiency of approximately 35 percent (Fig. 6.3).

More important than simple efficiency is the figure of merit (FOM), measured in dollars per watt per kilogram. In this FOM the spectrally split rainbow technique is calculated to attain close to a factor of two improvement over vertically stacked cells (both with four cell types) and considerably more with a larger number of cell types, and a factor of four improvement over silicon planar arrays. Some of this improvement comes not from the efficiency gain but from the ability to use

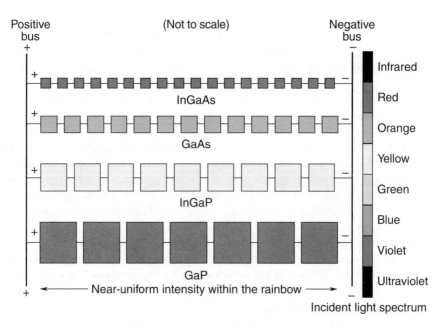

Fig. 6.2. Voltage matching layout.

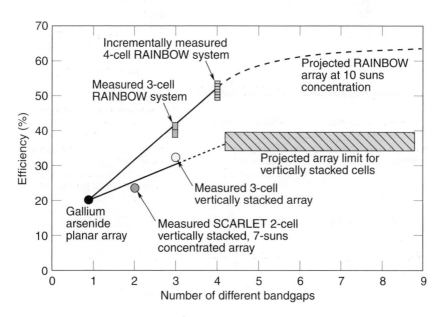

Fig. 6.3. Solar array efficiency.

higher concentration ratios due to the better thermal rejection of this type of array. Solar power conversion figures of merit for three options are illustrated in Fig. 6.4.

The high conversion efficiencies attained with these solar arrays, as well as development of solar dynamic converters and even nuclear sources, will encourage near and midterm fielding of high power spacecraft. A midterm driver will be power beaming from central power sources in space to constellations of using spacecraft. A longer-term driver will be delivery of massive quantities of space solar power to Earth power grids via power beaming. Regardless of the means of generating the power, a major problem will be the ability to efficiently reject large amounts of heat in space. Conventional surface radiators will be large and heavy in such very high power applications. A solution will be to implement a lightweight, high capacity heat radiator, a known but undeveloped technique that gains a fundamental advantage by using small dust or metal particles as heat carriers.

The technique, illustrated in Fig. 6.5, makes use of the large surface area-to-mass ratios of small spheres. It transfers source heat to a stream of such particles, ejects them into space where they cool rapidly by radiation, and collects the cooled particles to be recycled. The weight and overall size of such a radiator will be much lower

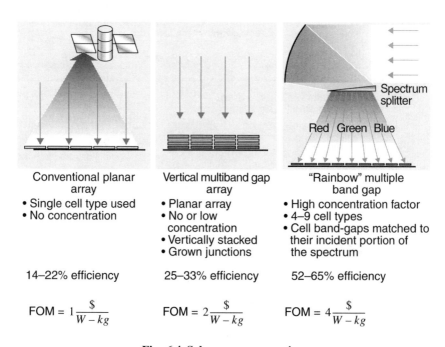

Conventional planar array	Vertical multiband gap array	"Rainbow" multiple band gap
• Single cell type used • No concentration	• Planar array • No or low concentration • Vertically stacked • Grown junctions	• High concentration factor • 4–9 cell types • Cell band-gaps matched to their incident portion of the spectrum
14–22% efficiency	25–33% efficiency	52–65% efficiency
$FOM = 1\dfrac{\$}{W-kg}$	$FOM = 2\dfrac{\$}{W-kg}$	$FOM = 4\dfrac{\$}{W-kg}$

Fig. 6.4. Solar power conversion.

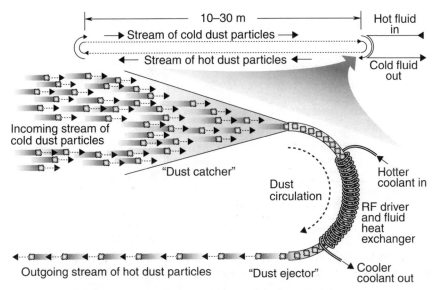

- Spherical dust particles are ferromagnetic material.
- Small spheres have large area-to-mass ratio.
- Heat is radiated away much faster than with sheet radiators.

Fig. 6.5. Lightweight, very high capacity heat radiator.

than a conventional solid area sheet radiator. Such radiators will become advantageous when tens of megawatts or more of power are being generated, with a sizable portion required to be rejected as heat.

A particularly useful combination of techniques will result in the development of megawatt-level modules for space use. Thin film optical concentrators, spectrally split and bandgap matched rainbow solar arrays, and dust particle radiators can be combined to implement 1-megawatt, high power modules, as illustrated Fig. 6.6. These modules can become the building blocks of a variety of future high power space systems.

Once the technology is demonstrated by the government for government uses, these modules are very likely to be developed by the commercial sector to produce and operate very large spacecraft whose function would be to sell power to energy-deficient nations. The attractiveness of such space solar power spacecraft is the clean and inexhaustible source of energy they represent. Space solar power satellites have been known since 1958 when conceived by Dr. Peter Glaser, but until very recently they were projected to require hundreds of billions of dollars to develop and even then could not deliver power at prices anywhere near competitive with conventional sources.

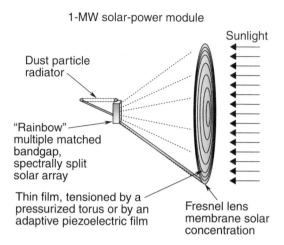

1-MW solar-power module

Sunlight

Dust particle radiator

"Rainbow" multiple matched bandgap, spectrally split solar array

Thin film, tensioned by a pressurized torus or by an adaptive piezoelectric film

Fresnel lens membrane solar concentration

Fig. 6.6. One-megawatt solar power module: 40-meter diameter, $20/watt, 500 watts/ kilogram, 2000 kilograms, approximate total cost $20 million.

All this is changing as a result of the publication of the NASA "Fresh Look" study of such solar power satellites,[6.2] which reexamined the satellites in light of modern technology and different architectures. The results indicate that the capital costs for a large solar power satellite are reduced by at least an order of magnitude compared with previous estimates if the satellite is made of replicated modules, such as the megawatt modules described above, rather than building a unitary satellite. That study estimates that these technologies, combined with benefits of mass production, will result in a module that produces 1 megawatt of solar power, weighs 2000 kilograms, and costs only about $20 million.

These megawatt modules will be designed to be combined to form space systems capable of generating hundreds of megawatts and measuring 10 kilometers along one dimension, converting the power into microwave or millimeter waves, and beaming it to the ground by kilometer-sized antennas. There the power would be received and rectified in rectennas 5 kilometers across and placed on a ground power grid. (See Fig. 6.7.) Beaming via lasers is also feasible and would require substantially smaller lasers and arrays of photocells. Ability to beam the power to rectennas anywhere makes such concepts ideal for delivering massive amounts of energy to developing nations, which is necessary if they are to advance to industrialized status. Such solar power spacecraft will be an indispensable contributor to the industrialization of space, to be discussed in Part III.

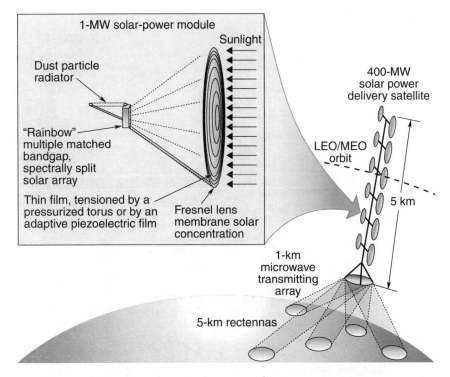

Fig. 6.7. Commercial delivery of high power from space.

The ability to field large reflectors in space also enables the fielding of large RF reflectors for beamed microwave or millimeter-wave power. Such reflectors will allow the wireless transmission of large amounts of electrical power from energy rich regions to those who need the energy, and to do so less expensively than using ground transmission lines. Since the relaying of power via space does not require the technology for power generation or radiation, it will be less expensive and will enable economically viable enterprises long before the space solar power satellites do so.[6.3] The likely sequence of development of such commercial power systems is shown in Fig. 6.8, beginning with the less demanding and cheaper LEO power relay systems, progressing through LEO solar power systems, and eventually resulting in GEO solar power delivery satellites. Capabilities such as these will be described more fully in a number of specific applications concepts in Part III.

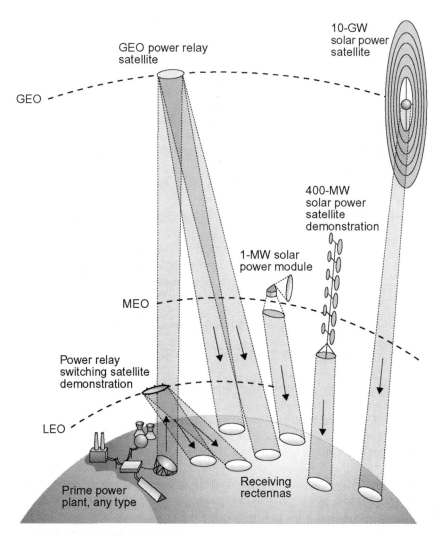

- Relay and source spacecraft
  - Relay tests phased reflectarray
  - Source tests phased array transmitters and high efficiency solar energy conversion
  - Both will be highly modular for mass production, low cost systems
- Modular phased arrays of 30–300 m
- 1–10,000 MW modular power units
- Can form components of power-intensive defense or civilian systems

Fig. 6.8. Commercial large-scale energy delivery likely development sequence.

6.1 References

[6.1] Ivan Bekey, "Rainbow's Array of Promises," *Aerospace America*, American Institute of Aeronautics and Astronautics (January 1999).

[6.2] "Space Solar Power: A Fresh Look at the Feasibility of Generating Solar Power in Space for Use on Earth," SAIC Report No. SAIC-97/1005 to NASA (April 4, 1997).

[6.3] Ivan Bekey (Bekey Design, Inc.), "An Innovative Multi-User Passive Space Power Relay System," *49th International Astronautics Congress* (Melbourne, Australia, October 2, 1998). Paper No. IAF-98-R.2.06.

7 Space Transportation

Space transportation is the most crucial and costly aspect of space programs. It will be treated in two parts: transportation within space and transportation from Earth to orbit.

7.1 In-Space Transportation

Moving and maneuvering within space, at LEO or beyond, is considerably easier than traveling from Earth to space, which calls for climbing rapidly out of Earth's gravity well to use minimal energy (to avoid the so-called gravity losses). Such launch requires high thrust engines, which are currently difficult to make with high specific impulse capability. Once the spacecraft is in orbit, much longer time can be taken to perform maneuvers with little energy loss, and low thrust engines can be used. They are much more likely to have high specific impulse, and therefore can impart orbital velocity changes with much less propellant than chemical high thrust Earth-LEO engines.

A number of existing in-space transportation techniques will eventually be augmented by others that have been identified but not yet developed. Solar sails, for example, have great promise. The principle has been known for decades but not yet applied in operational spacecraft, except as small torque-producing surfaces for attitude control of a few planetary spacecraft. The sails produce thrust by reaction of solar photons and can be made to literally "sail" in space, as the thrust vector is normal to the film surface. The ability to manufacture large thin films in space will enable lightweight solar sails to encompass large areas and thus to generate much greater accelerating force from sunlight photon pressure than sails using present-day films. A second technique using sails bends the direction of particles of the solar wind using a magnetic field generated by a large superconducting coil circulating a very large current. These sails are illustrated in Fig. 7.1.

Increasing the acceleration capability of photon sails requires a decrease in mass, which is achieved by using thinner materials for the film while maintaining large surface area. The difficulty of deploying the sails without tearing the material limits the attainable acceleration. Fig. 7.2 illustrates a solution using the benign space environment to fabricate such sails in space. The film can be spun from a folded configuration or even from a partially molten mass by slowly spinning the mass into a large, thin, rotating disk. Radial and circumferential tethers would be embedded to hold the film against tearing and would weigh very little because they are tension-only structures subjected to low forces. Once deployed or manufactured in situ, the plastic film would be aluminized by a travelling vapor-deposition robotic device maneuvering in close proximity. Once aluminized, the sail would be exposed to sunlight on its plastic side, and the material would evaporate or sublimate in time. This would leave a rotating aluminum film only a few atoms thick, supported by the tethers. Additional tethers would lead from these supports to form a harness that would pull the payload at some distance

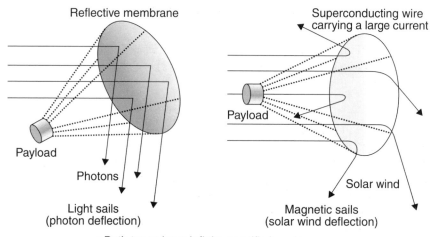

• Both types have infinite specific impulse
• Can exit solar system at hundreds of km/sec
• New materials warrant reexamming sails for:
  – planetary exploration
  – small-body missions
  – Earth-orbit applications

Fig. 7.1. Solar sails.

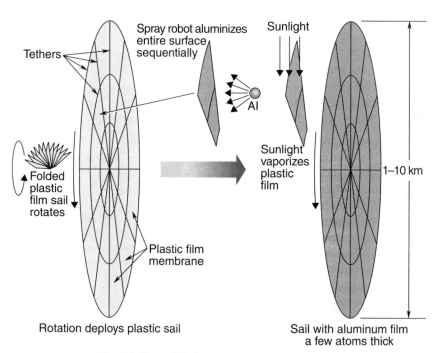

Fig. 7.2. Space fabrication of super-light solar sail.

from the sail. Sails could be manufactured in space to be tens of kilometers across, yet light enough to generate an acceleration of 0.01 gravity at a distance of 1 astronomical unit (AU) from the sun, which is one to two orders of magnitude greater than possible using current techniques that manufacture and fold the sail on the ground. In addition, the sails will eventually be manufactured wholly from Buckytubes, which will further increase their acceleration capability to the order of 0.1–1 g at 1 AU and more nearer to the sun. Buckytubes will be discussed in Chapter 9.

Sails of this type will be capable of relatively rapid maneuvering in space; they will have infinite specific impulse, requiring no propellants. They are reusable in the fullest sense of the word and simple enough to have a long life operating in space. They will be useful for Earth orbit transfers, but only above about 1000-kilometer altitude because of their extreme sensitivity to atmospheric drag. They will be particularly useful for propulsion throughout the solar system, including outer planet missions where their sizes can simply be increased to overcome the loss of solar energy because of their distance from the sun.

Solar sails are but one of many advances in transportation within space that will make their appearance in the 30-year time frame of this study. The advent of large solar power collection and electrical power generation, to be discussed in Part III, will enable ion propulsion at megawatt power levels. This means that a thrust of 1000 Newtons could be attained from a 25-megawatt assembly identical to that required for a space solar power system but without the microwave generator or antenna. Such systems will become high thrust, high specific impulse transfer vehicles for robotic and human planetary exploration.[7.1]

It is clear, as it has been for many years, that nuclear-powered stages are most desirable because of their simultaneous high specific impulse and high thrust. A number of designs for stages incorporating a variety of reactor designs have been made, but none have so far made it to the flight stage. Various fusion-powered stage designs that would trade more benign environmental operation for great size and weight have also been studied and proposed. Deep space applications using fusion and power beaming will be treated further in Part III.

The ultimate technique for nuclear propulsion, understood in principle only, is matter-antimatter annihilation, capable of at least 1000 times the energy density of nuclear reactions (which makes them a billion times greater than chemical energy density). The practical implementation of such matter-antimatter annihilation systems is probably at or beyond the 30-year horizon of this study. Nonetheless they are likely to make their appearance, given that antimatter is being produced and stored today, although in minuscule quantities. Notwithstanding their obvious benefits, all nuclear systems appear to have political drawbacks.

Thus a large number of conventional and unconventional in-space transportation techniques will become practical and available by 2030–2040. Some of these are illustrated in Fig. 7.3.

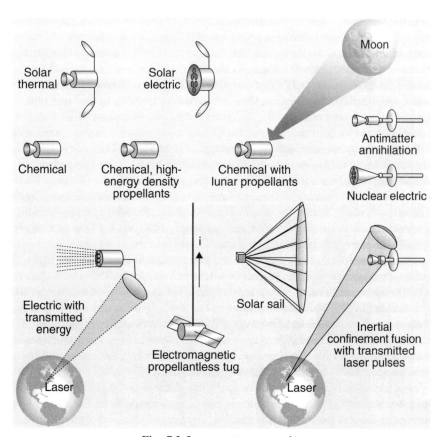

Fig. 7.3. In-space transportation

7.2 Earth-to-Orbit Transportation

Launch vehicle costs (prices) in dollars per kilogram to orbit have been relatively static for many years, compared with the dollars per kilogram of many types of spacecraft, which have been steadily decreasing as a result of new technologies and manufacturing methods. Launch costs are from one-third to one-half the total cost of fielding most space systems today. Launch vehicles are fragile, complex, 1000 times less reliable and 10,000 times more expensive than airliners per kilogram of payload. Although launch vehicles are functional, as evidenced by their regular use to loft spacecraft, the operational and economic constraints of their characteristics are the largest obstacles to more widespread and routine use of space. That is about to change dramatically.

This change will not come from the advent of the Evolved Expendable Launch Vehicle (EELV) systems now in development. Expected cost savings for EELV over current vehicles are no more than about 25 percent because of infrequent launches—approximately one per week, costs for the expended hardware which consists of the entire vehicle, and labor to assemble and validate what is essentially a new vehicle every launch. At best, these launch vehicles will result in launch costs of $15,000 per kilogram and reliabilities of 0.99, just matching those of the Space Shuttle, which is the most reliable launch vehicle ever flown.

This way of operating diametrically contrasts with that of airlines, which daily "launch" tens of thousands of aircraft that are reusable thousands of times, expend nothing but propellants, operate at $10 per kilogram of payload, and maintain reliabilities of 0.999999. A number of technologies are being investigated even today that in time will allow space launch vehicles to operate in the same manner as airliners, at as low a cost per flight, if not lower. This is not a pipe dream, and it is not a question of "if," but rather of "when." The technologies have the potential to reduce the cost of Earth-to-orbit space launch by about one order of magnitude per decade until costs per kilogram of payload reach levels similar to those experienced by airliners today. Each of these technologies is being investigated by NASA and could result in the development sequence shown in Fig. 7.4, which is definitely attainable, given the will and funds.

The first step will see the expendable vehicles such as Atlas, Titan, Delta, and the EELVs, and the partially reusable Space Shuttle, replaced by fully reusable vehicles. The Reusable Launch Vehicle program (RLV) by NASA and Lockheed Martin was to develop a fully reusable launcher in the 2005 time period, the Venture Star, whose operating costs were designed to be $2000 per kilogram of payload, which is an order of magnitude lower than that of current vehicles. A half-scale experimental vehicle, the X-33, was to demonstrate the technology. The design would be for a single-stage-to-orbit vehicle, the class of rocket vehicles with the lowest possible operating costs. A set of problems and funding limits resulted in premature abandonment of this worthy and attainable goal early in 2001. It was replaced by a set of yet-to-be-defined, second-generation partially or fully reusable vehicles—probably two-stage—to replace the Space Shuttle in 2012. Although none hold the promise of the Venture Star, nonetheless some might operate at costs half an order of magnitude lower than current vehicles, to $6000 per kilogram, and be operating during the next, or at most two, decades. Purely commercial fully reusable two-stage vehicles, such as the Kistler, could also be developed in the early part of this time frame and would operate at about $4000 per kilogram of payload.

The next major launch cost reduction could occur in two to three decades in one of two ways. The first is the use of more advanced technologies in fully reusable single-stage-to-orbit vehicles. Examples, among others, are
- more advanced propulsion utilizing air breathing in the very early parts of an accelerator trajectory and rocket propulsion the rest of the way
- higher energy density propellants
- fully automated checkout, launch, and fault detection systems
- maglev catapult for launch

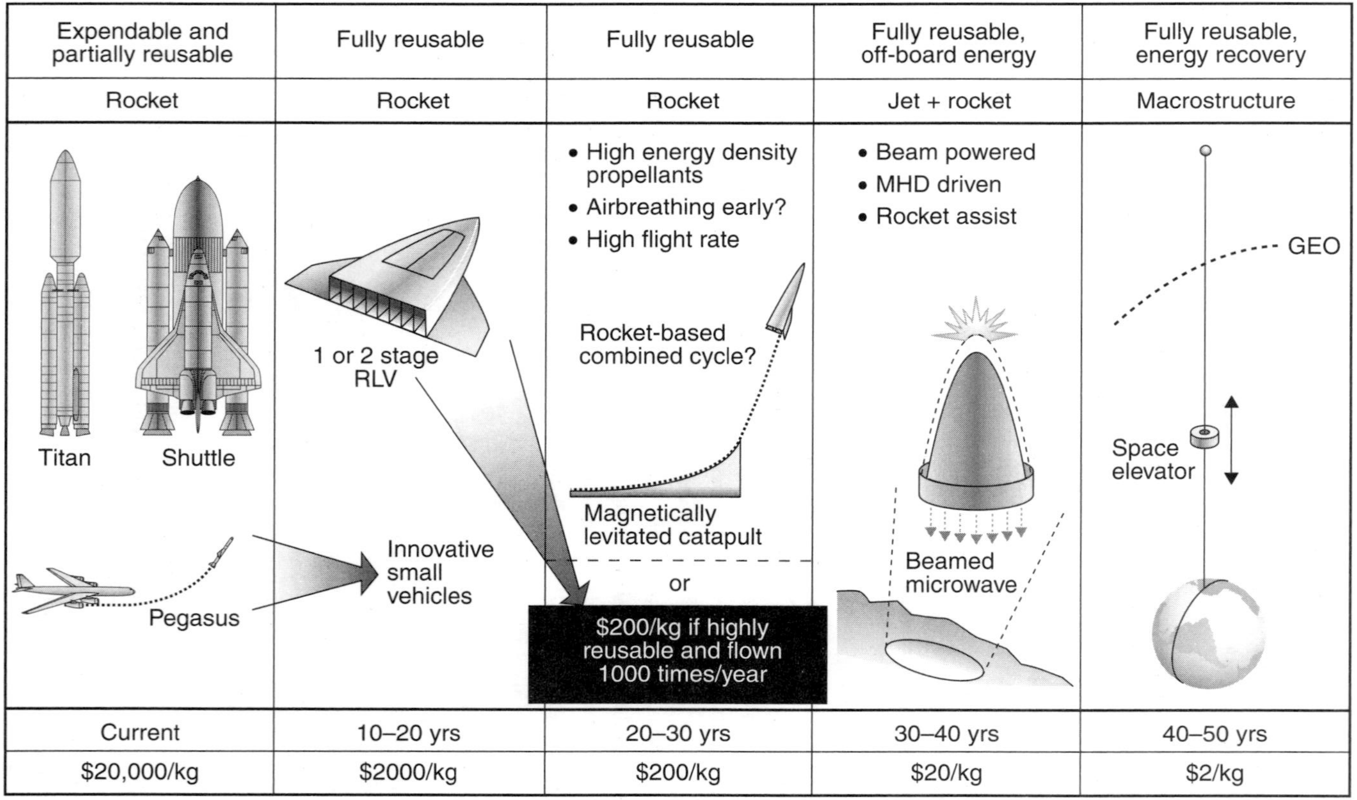

Expendable and partially reusable	Fully reusable	Fully reusable	Fully reusable, off-board energy	Fully reusable, energy recovery
Rocket	Rocket	Rocket	Jet + rocket	Macrostructure
Titan Shuttle Pegasus Innovative small vehicles	1 or 2 stage RLV	• High energy density propellants • Airbreathing early? • High flight rate Rocket-based combined cycle? Magnetically levitated catapult or $200/kg if highly reusable and flown 1000 times/year	• Beam powered • MHD driven • Rocket assist Beamed microwave	GEO Space elevator
Current	10–20 yrs	20–30 yrs	30–40 yrs	40–50 yrs
$20,000/kg	$2000/kg	$200/kg	$20/kg	$2/kg

Fig. 7.4. Launch into space. (MHD—magnetohydrodynamic)

These will allow operation at greater margins, high reusabilities, and low maintenance, which will bring about much lower costs. They can reduce the costs of launch by another order of magnitude, to the order of $200 per kilogram. NASA is using this approach in its third-generation, reusable space transportation program.

A space mass market, which will require thousands of launches annually, is the second way to bring about a similar decrease in launch costs. A number of potential commercial space applications will be large enough to require, and even greatly exceed, this level of launch activity. In fact, some will be so large as to dwarf today's combined government and commercial space activity. Under these conditions, reusable launch vehicles without the advanced technologies described above, but oversized and designed with automation and much greater margins making them capable of very many reuses and autonomous operations, will also attain the second order of magnitude cost reduction without further technological advances.

This point is illustrated in Fig. 7.5, which is a plot of several fully reusable launch vehicle designs based on data by Jay Penn, of The Aerospace Corporation. Each point represents a different fully reusable single-stage-to-orbit vehicle, each designed for a greater number of flights before refurbishment of the vehicle and the

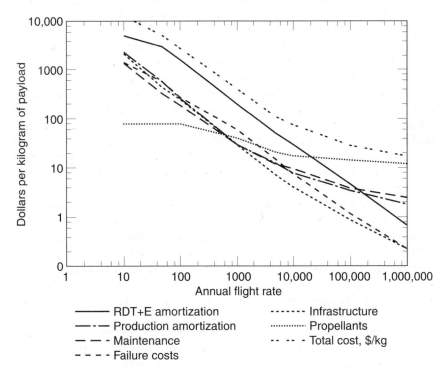

Fig. 7.5. Flight rate effect on launch costs. (RDT&E—Research, Development, Test & Engineering)

the engines. These greater number of flights raise the research and development and fleet production costs, which must be amortized among the number of flights. However, they result in a much greater savings in operations cost than the research and development amortization increase. The curves show that if a market can be found that requires a large number of launches per year, chemically propelled single-stage-to-orbit reusable launch vehicles can operate at total costs approaching $20 per kilogram of payload. In fact, they might cost even less because that number was limited by overly conservative estimates for the minimum cost of liquid hydrogen fuel in high volume production.

This begs the question: Where is the market that needs thousands of flights per year? The answer lies in unconventional commercial activities in space that will grow to be staggeringly large if the cost of launch is sufficiently reduced. This was the conclusion of the Commercial Space Transportation Study, done in the early 1990s by a consortium of the largest U.S. aerospace companies. In fact, the orbiting of tens of millions of kilograms of space solar power satellites will require thousands of launches annually, even with new large vehicles. So will the frequent passenger launches that will be demanded for public space access (space tourism) as the businesses mature. These and other nontraditional commercial markets could well drive launch costs down to several tens of dollars per kilogram of payload simply due to their large market demand for launches. Such low costs would have a tremendously beneficial impact on all space activities.

NASA and the Space Transportation Association have determined in a cooperative study and workshop[7.2] that public space access will be real and of enormous magnitude soon after low cost, highly reliable launch vehicles are available. A companion study by the author[7.3] determined the clear economic viability of many business ventures dedicated to transporting persons into space, and forecast potential annual revenues in the tens of billions of dollars for such a new space market. This companion study determined that potential businesses offering such services could make internal rates of return in the 35 to 55 percent range by developing and operating reusable launch vehicles using essentially current technologies but designed for heavy reuse.

Thus it is clear that both the market and the means will exist to drive down the costs of launch to orbit to well below $200 per kilogram and in all probability close to $20 per kilogram. In fact a number of entrepreneurial ventures are already under way to provide tourists with suborbital space travel and will grow into orbital tourism in the next decade using current technology vehicles. Paying passengers have already been to the Russian space station MIR, paying a reported $25 million, and to the International Space Station, paying a reported $20 million, and others are sure to follow.

Fig. 7.6 shows the results of a number of market surveys in this area and demonstrates the elasticity of this market. The average of the data indicate that even at $1 million per ticket, up to 100 tourists annually will be willing to go into space. When the launch costs drop sufficiently and result in ticket prices of $100,000,

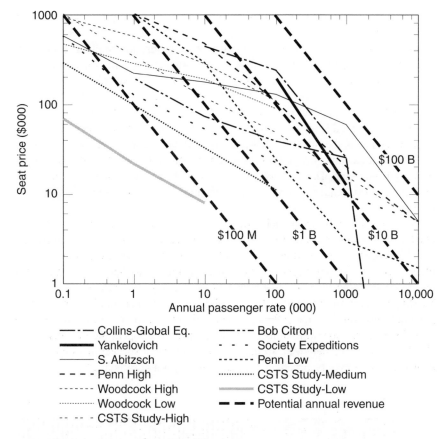

Fig. 7.6. Market elasticity of public space travel and potential annual revenues.

100,000 persons will want to go into space each year. And when further cost reductions result in ticket price drops to $20,000, more than 1 million tourists annually will want to go into space. The figure also shows that the market is tens of billions of dollars per year, which virtually guarantees the appearance of entrepreneurs.

Studies have shown that the low ticket prices required for this size market to develop are achievable with properly designed vehicles and businesses. The needed drop in launch vehicle costs by two orders of magnitude can occur if the new markets develop and launch vehicles are designed using more advanced technology and conservative features for long operating life. But somewhat of a circular dilemma exists in this development because the existence of the markets is needed to obtain funding to develop the vehicles and start the businesses, but the vehicles must be developed and proven before the markets themselves can obtain the funding to become real and credible. In practice, markets and vehicles will probably develop incrementally in parallel. Thus in 20 to 30 years, we should see at best launch costs of $200 per kilogram, and at least many hundreds to thousands of flights annually.

An entirely different technology will also make its appearance in this time frame and may be capable of low launch costs. Surprisingly, it will consist of mostly expendable systems. It is a variant of artillery guns, and is illustrated in Fig. 7.7. It is a Russian idea, considered by NASA,[7.4] which lines the barrel of a gun with a large number of explosive rings. Alternatively, the barrel could be eliminated, with the rings suspended from a simple rail. When the explosive rings are detonated in precise sequence, the high and constant pressure behind an encapsulated payload accelerates it extremely rapidly, so it is capable of exiting at Mach 27 from a gun only 30 meters long. This technology makes possible direct launch to an apogee at orbital altitudes, requiring only a circularization stage to function as a spacecraft launcher.

The high acceleration, 100,000–300,000 g's, limits its use to dense, well-packaged spacecraft or commodities such as raw materials, propellants, or water for in-orbit propellant production via electrolysis. This blast-wave accelerator has so little explosives, fixed structure, and infrastructure that their replacement and refurbishment after the explosive launch is estimated to be in the range of $200–$2000 per kilogram of payload. This accelerator can also be an ideal launcher for placing many small spacecraft or modules of spacecraft into orbit, there to be joined into larger operating units or to be operated as cooperating swarms. It can also be made in smaller sizes, so that efficient launch of spacecraft weighing less than 1 kilogram may be practical, though probably not at the same low cost.

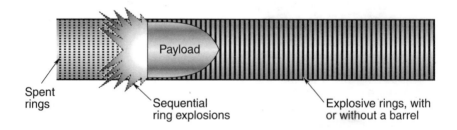

Spent rings

Sequential ring explosions

Explosive rings, with or without a barrel

- Blast wave accelerator
- Series of explosive rings
- Rings detonated in precise sequence
- Essentially constant acceleration
- 15-m barrel results in orbital velocity at 300,000 g's
- 40-m barrel reduces to 100,000 gravity
- Russian experiments indicate Mach 27 achievable
- Suitable for dense payloads or commodities

Fig. 7.7. Ground-based blast wave acceleration for use as orbital launcher or global range gun.

The third major step in reduction of space launch costs will require elimination of most propellants on the launch vehicles. This step envisions placing the energy source on the ground and beaming the energy to the vehicle while using as much free air as possible for the oxidizer and reaction mass. Concepts are already under investigation for such a vehicle, which uses millimeter waves as the received energy to break down air in front of it and accelerates the resulting ionized air by MHD forces, using superconducting coils near its base. The vehicle could hover or accelerate and would use oxygen and hydrogen from small internal tanks as a rocket for the final ascent to orbit. In principle, this would be even more efficient than a ground-based laser, which is used to ablate a plate at the bottom of the vehicle, which requires reaction mass because it is not supplied by the atmosphere.

This technique uses electrical off-board energy with costs on the order of $0.10 per kilowatt hour, which translates to less than $2 per kilogram of payload. However, because of the necessary onboard final impulse rocket mass and operations costs, the complete system cost will probably be closer to $20 per kilogram of payload. This kind of a vehicle can also hover in the atmosphere, being a vertical jet, and could be used for powered descent and landing as well as launch, making it well suited for reusable single stage vehicles. It could be practical within 30 to 40 years, and would reach the typical operating costs of commercial airliners today.

While $20 per kilogram of payload is extremely impressive compared with today's launch costs that are 1000 times greater, that is not the end of the road. A number of macrostructure concepts have intrinsically lower operating costs. An elevator riding on a tether connecting the surface of the Earth to a counterweight at supersynchronous altitude is one example. Since a point in GEO hovers over one point on the equator, in principle a tether could be extended between them. The counterweight for this system would be on an extension of the tether beyond GEO. The structure, once installed, is passive and stable and can be reused indefinitely. The elevator would be electrically powered and would carry the payload, which would be simply released when at GEO altitude.

The materials for this tether elevator structure are already being produced in laboratories, and are addressed in Chapter 9. Their high strength-to-weight ratio and practicality will surprise many. In fact, such elevators have been the subject of recent NASA studies.[7.5] An electrically driven elevator can ride up and release a payload in GEO for a total cost of about $10 per kilogram, which is already lower than commercial aircraft operating costs today. However, if the elevator returned the payload (or a previously orbited payload with the same mass) to the surface for servicing or upgrading, the energy generated by electrodynamic braking during descent would be almost exactly equal to that expended in lifting it to GEO. Thus, in principle, if the average mass of a number of spacecraft lifted to GEO in a given time is equal to that of spacecraft returning to the Earth's surface, on the average the system could operate requiring no net energy, except to make up small losses. Therefore, in practice it should be possible to reduce launch costs to about $2 per kilogram of payload placed into GEO. This would be a reduction of launch costs of four orders of magnitude from today's levels.

In conclusion, with the expected reduction in launch costs, which could be as high as one order of magnitude reduction per decade for four decades, launch into space will be inexpensive indeed and more similar to commercial aircraft operating costs today. This should not be surprising, since the energy required for an orbital launch and a transatlantic airliner flight are within an order of magnitude, and if no hardware is expended, crew sizes are similar, and the space vehicle is powered by electricity, the spaceliner and airliner payload operating costs should also be similar.

It seems clear that in the next 20 to 40 years, space will become a very different place. Access to it will be no more expensive than today's transatlantic or transcontinental airline flights. Furthermore, the spacecraft to be launched will be large but lightweight, even with antenna and optics the sizes of hundreds of meters. Further major reductions in weight of spacecraft using new materials in the offing are discussed in Chapter 9.

7.3 References

[7.1] John C. Mankins, "Affordable Mars Exploration Architectures—Applying Systems from the Commercial Development of Space," *50th Congress of the International Astronautical Federation* (Amsterdam, the Netherlands, October 4–8, 1999). Paper No. IAA-99-IAA.13.3.05.

[7.2] Daniel O'Neil, Ivan Bekey, John Mankins, Thomas F. Rogers, and Eric Stallmer, "General Public Space Transportation and Tourism," *Executive Summary*, vol. 1, NASA Marshall Space Flight Center Report no. NP-1998-03-11-MAFC (March 1998).

[7.3] Ivan Bekey (Bekey Designs, Inc.), "Economically Viable Public Space Transportation," *49th International Astronautics Congress* (Melbourne, Australia, September 28–October 2, 1998). Paper No. IAA-908-IAA.1.5.05.

[7.4] D. E. Wilson, Z. Tan, and P. L. Varghese, "Numerical Simulations of the Blast-Wave Accelerator," *AIAA Journal*, Vol. 34, No. 7, pp. 1341–1347 (1996).

[7.5] Bradley C. Edwards, "The Space Elevator," NASA Institute for Advanced Concepts (2000), http://www.niac.usra.edu/studies.

8 Spacecraft and Spacecraft Busses

Spacecraft and their busses will also change dramatically over the longer term. In fact, the spacecraft bus as we know it today will essentially disappear because spacecraft will be either very large and of gossamer materials and use information and control instead of structures or will be very small, produced in a foundry by the hundreds of thousands and emplaced into cooperating swarms. Neither implementation lends itself to today's typical two-element spacecraft—the bus and the payload. In fact in both the above examples the housekeeping functions are integrated with the payload functions and both are distributed.

The advent of very much lower launch costs will bring another trend. The current appeal of so-called microsatellites, weighing a hundred kilograms or so, will come into serious question. If launch costs drop to almost nothing, why build small and lightweight spacecraft? Furthermore, if large spacecraft can be constructed from thin films or new lightweight materials, the spacecraft as well as its launch should be very inexpensive even if the spacecraft is very large, and the current necessity of miniaturization for cost reduction will disappear.

It seems clear that when launch becomes very cheap, today's aerospace culture, which strives to minimize the weight of every component and imposes intensive testing and careful quality control because long life is a necessity, could change radically. This is illustrated in Fig. 8.1. It is this culture and set of labor-intensive practices that drive spacecraft costs as high as they are; it is not the cost of the materials used. For instance, why build spacecraft and component structures as light as possible, making them so thin they just barely survive under the launch loads, and require extensive testing and integration to insure such survival? These are people-intensive operations and are the dominant component of spacecraft costs.

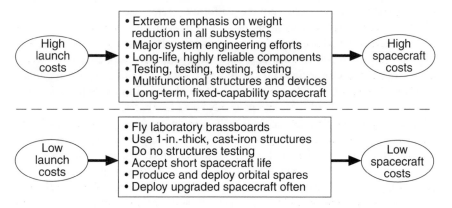

Fig. 8.1. Impact of low launch costs on spacecraft costs.

For example, if a launch costs virtually nothing, why not build the spacecraft structure out of 3-centimeter-thick cast iron, with hardly any design effort, and never test it at all? That would surely be much cheaper, and with a little judgment could be every bit as good and practical. So what if it added 1000 kilograms to the spacecraft weight if the cost to launch that extra weight were only $20,000 (at $20 per kilogram)? While this example is admittedly simplistic and not necessarily proposed, today's culture that calls for spacecraft design, construction, and testing to minimize weight and maximize reliability will have to be reexamined when launch becomes essentially free because many operations can be simplified or eliminated if weight is not a concern. Because of the low launch costs, a number of generations of laboratory brassboards could also be orbited to complete the development process of a space system in orbit, its operating medium. Thus equal if not higher confidence would be gained such that the final spacecraft would have the desired reliability, yet it would have been developed and produced at lower cost.

In addition, these very low launch costs will mean that it will be economically feasible to service and upgrade many types of spacecraft and space systems in orbit. Thus it may become attractive to launch spares or to service a spacecraft in orbit, much as we service ground and airborne equipment, rather than depending on it to function without maintenance for many years. The establishment of the service life of a spacecraft will then become a business, rather than a technical, decision, as it should be.

Notwithstanding the above trend, some spacecraft busses will still be used but most will likely be larger multifunction platforms, serving a number of users or payloads that could rent the housekeeping functions from the platform operator. These platforms will allow new space functions to be performed without the payloads or "front ends" having to provide for propulsion, attitude control, translation control, solar arrays, thermal rejection, or other "bus" functions. Instead, new spacecraft consisting only of the true payload components will be docked into one of several payload ports on the platforms, and the payload will purchase housekeeping and support services from the platform operator. (See Fig. 8.2.)

Initially, automated spacecraft servicers will be designed to dock with a spacecraft to exchange payload modules as well as bus subsystem modules. These will be plug-and-play modular constructions, with separate fiber-optic data and power busses. Increasing crowding in GEO will give rise to such multipurpose, multiuser GEO platforms, operated by the private sector, which will provide housekeeping functions for a variety of front-end payloads on a lease basis.

Eventually a new class of servicing will become practical, in which small servicers dock with spacecraft or platforms in orbit at locations designed to provide access to their internal data and power networks. Once docked, a microservicer would replace a failed function through the data and communications bus without necessarily having to replace the failed part. Picosatellites could efficiently perform this service and would handle many, though not all, failure types. These capabilities will naturally engender the growth and adoption of interoperability standards because of the economic advantage they would afford.

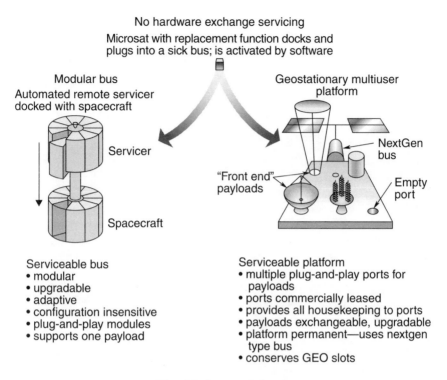

No hardware exchange servicing
Microsat with replacement function docks and
plugs into a sick bus; is activated by software

Modular bus
Automated remote servicer
docked with spacecraft

Geostationary multiuser
platform

Servicer

NextGen
bus

"Front end"
payloads

Empty
port

Spacecraft

Serviceable bus
• modular
• upgradable
• adaptive
• configuration insensitive
• plug-and-play modules
• supports one payload

Serviceable platform
• multiple plug-and-play ports for
  payloads
• ports commercially leased
• provides all housekeeping to ports
• payloads exchangeable, upgradable
• platform permanent—uses nextgen
  type bus
• conserves GEO slots

Fig. 8.2. Spacecraft busses

In the longer term, adopting these innovations will result in spacecraft very different from today's bus-and-payload designs. These spacecraft will be composed on the one hand of self-contained but cooperating swarms of picosatellites and on the other of very large gossamer film ensembles, both of which use information and control in place of structures and which have tethers, station-kept elements, and distributed housekeeping functions. Their weights and reliabilities will also be radically different, as decisions regarding service lives and in-orbit servicing and upgrading become routine and business based. In the aggregate these trends will greatly diminish the importance of weight as a consideration in development and launch of space systems, and will make space systems very inexpensive compared to costs today.

9 Revolutionary Materials and Structures

Preceding chapters introduced technologies that will work to make weight much less significant in future space systems. This chapter discusses new materials with radically greater strength/weight ratios that will have truly revolutionary effects on spacecraft weight, far beyond even dramatic reductions already presented.

An entirely new class of structural materials based on pure carbon nanotubes, a variant of the so-called "Buckyballs," already exists in the laboratory and will become widely available during the 2010–2030 time period. Carbon atoms arranged into long thin hollow tubes—nanotubes—are known as "Buckytubes." Their diameter is in the order of a nanometer, approximately 1/2000 of the wavelength of light, hence their name. They have been grown in the laboratory in micron lengths to date, although in principle there is no limit to their eventual length. (See Fig. 9.1.)

Their overriding advantage derives from their pure carbon-carbon atomic bonds, which, arranged in a lattice like a rolled sheet without defects, are the strongest and stiffest in nature and result in a strength-to-weight ratio 600 times greater than that of high strength steel or aluminum alloy. This phenomenal strength has been verified experimentally. Thus they are comparable to the previously unobtainable and only postulated highest strength material: long crystals of pure diamond. Their physical characteristics are shown in Table 9.1.

Buckytubes can be thought of as analogous to the carbon fibers used in carbon-composite structures, except that they are pure molecular structures that lack the imperfections that limit the strength of today's carbon fibers. They are already being made in a number of laboratories in gram quantities in the United States and abroad, such as by researchers at Rice University under Dr. Richard Smalley. Carbon Nanotechnologies, Inc., under Dr. Ken Smith, operates a pilot plant capable of making half a kilogram of pure single-walled, though randomly oriented, carbon nanotubes

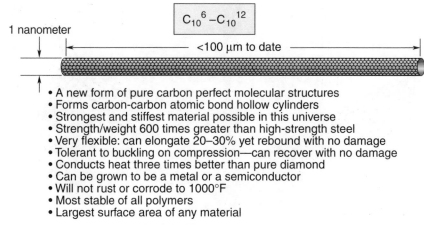

$$C_{10}{}^6 - C_{10}{}^{12}$$

1 nanometer

<100 μm to date

- A new form of pure carbon perfect molecular structures
- Forms carbon-carbon atomic bond hollow cylinders
- Strongest and stiffest material possible in this universe
- Strength/weight 600 times greater than high-strength steel
- Very flexible: can elongate 20–30% yet rebound with no damage
- Tolerant to buckling on compression—can recover with no damage
- Conducts heat three times better than pure diamond
- Can be grown to be a metal or a semiconductor
- Will not rust or corrode to 1000°F
- Most stable of all polymers
- Largest surface area of any material

Fig. 9.1. What are Buckytubes?

Table 9.1. Comparative Strength of Buckytubes

Material	Young's Modulus (GPa)	Tensile Strength (GPa)	Tensile Strength (lb/sq.in.)	Elongation at yield (%)	Density (Water=1)	Strength/ Weight (Steel=1)
Aluminum alloy	70	0.5	75,000	11.0	2.7	1
High-strength steel	200	1.4	210,000	0.7	7.8	1
Kevlar	120	3.0	450,000	2.5	1.4	12
Carbon fiber	300	4.0	590,000	1.3	1.5	15
Pure-carbon, single-wall nanotubes (Buckytubes)	900	200.0	29,000,000	20.0	1.6	600

per day using a carbon monoxide vapor catalysis mechanism and projects producing tons per day in 2004. Dr. Smalley's research is intended to eventually grow at least meter-long, oriented, single-wall Buckytube structures in net shapes from metallic seeds, making possible single crystal sheets, mats, panels, and a variety of structural members with complex shapes. The method is illustrated in Fig. 9.2.

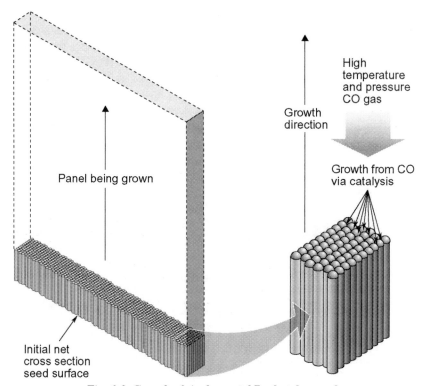

Fig. 9.2. Growth of single crystal Buckytube panel.

In contrast to composites, however, ultimately these materials will not require any matrix compound such as resin. Though their great strength is essentially uniaxial, van der Waals forces between nanotube sidewalls as well as growth of such panels along complementary directions will produce structures with great strength along many directions. Composites using matrix materials to bind Buckytubes will probably appear considerably earlier than the pure materials and will have strengths intermediate to current carbon-fiber composites and pure Buckytubes. The strength development expected for these materials is illustrated in Fig. 9.3.

These will be wondrous materials indeed. In addition to revolutionizing electronics, a broad topic recognized but not treated here because of time and resource limitations, their advent will revolutionize all structures—on the ground, in the air, and particularly in space. They will be used to make what is usually understood to be "structures"; however, they will also be used to make plumbing, pressure vessels, electronic circuit boards, container packages, batteries, antennas, deployment mechanisms, optics, wiring, rocket cases, and nozzles among other items. There is hardly any component of spacecraft that could not eventually be made of Buckytubes.

Because existing spacecraft components are not all sized by strength/weight considerations and utilize many materials, the use of Buckytubes will not have a uniform 600/1 spacecraft weight decrease. Some components such as pressure vessels and other tension-controlled members that are currently made of graphite composites will have weight decreases of a factor of 40–60. Aluminum panels booms, trusses, supports, and other components that are bending stress-controlled

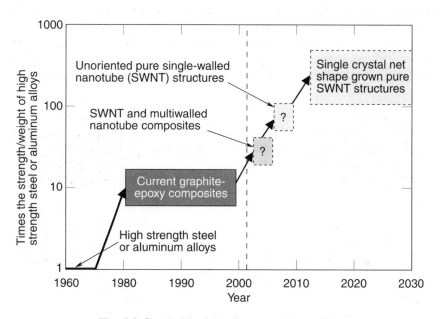

Fig. 9.3. Strength/weight of structural materials.

will have weight decreases by a factor of 20–30 if their linear dimensions must be maintained and Buckytube material is used to reduce their thicknesses. However, weight of these components will decrease by a factor of 400–600 if their linear dimensions are reduced while carrying the same load but their wall thicknesses are maintained constant. A reasonable estimate is that in the aggregate a typical spacecraft weight reduction factor of 100 can be expected if most components are made from Buckytubes and use clever structural design. The impact on spacecraft weight is nothing short of amazing, and is illustrated in Fig. 9.4.

Launch vehicles will also be made mostly from Buckytube materials. The effect of reducing the structure and other components, except the propellants, of a single-stage launch vehicle by a factor of 100 is to increase its payload to LEO by a factor of 10, with no size changes. Conversely the entire launch vehicle, including propellant tanks and engines, could be made 10 times smaller with no decrease in the vehicle's lift capabilities. This is dramatically illustrated in Fig. 9.5.

Even at "only" the projected 100/1 reduction of spacecraft weight and ignoring the 10/1 increase in launch vehicle capability, spacecraft will be so light while so large and capable that space systems will effectively weigh "nothing" compared with typical weights today. Terrestrial applications of Buckytubes will abound, exemplified by highly efficient automobiles light enough so that they could be carried to tight parking spaces by their owners. Single-stage launch vehicles will become eminently practical and lift 10 times their own dry weight; they will completely displace staged and expendable launch vehicles. Even the elevators from the surface to GEO altitude will become practical when made of Buckytubes, with the total weight of a tether elevator long enough to span that distance being 10,000 times less than if constructed from today's materials, and only about twice the weight of the payload it is to elevate. These examples are shown in Fig. 9.6. In fact, the launch costs discussed in Chapter 8, which were reduced to below $20–$200 per kilogram of payload by new technologies and frequent flights but using relatively

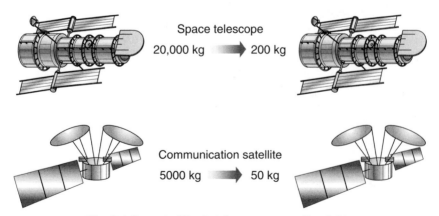

Fig. 9.4. Impact of Buckytubes on spacecraft weight.

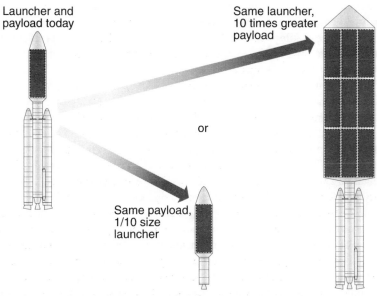

Fig. 9.5. Impact of Buckytubes on access to space.

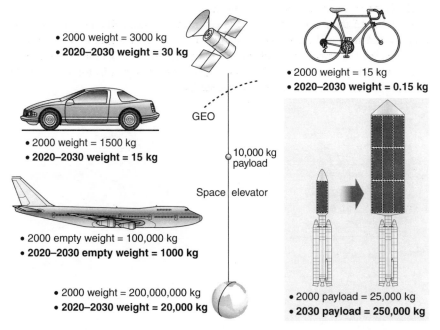

Fig. 9.6. Impact of Buckytube structural materials with strength/weight "only" 100 times that of high-strength steel.

conventional materials, could become up to an order of magnitude lower when the launch vehicles are constructed of Buckytubes if all effects were additive. Conversely the $20–$200 per kilogram cost levels will be attained at an order of magnitude lower flight rates and up to a decade earlier, thus enabling earlier, larger, and more diverse new businesses in space.

The combination of the up to 100/1 reduction of spacecraft weights for the same function and the up to 10/1 increase in the weight lifting capability of launch vehicles for the same size payloads results in a net reduction of effective space system weights of up to a factor of 1000. Because costs are roughly proportional to weights in all past and current space activities, it is likely, though not assured, that there will also be up to a thousand-fold reduction of space system development, production, and deployment costs. In principle, these reductions would be additive to the launch cost reductions of Chapter 7 engendered by technologies other than the use of Buckytube materials, which were up to four orders of magnitude in themselves, if the effects were linear and could be superimposed. However, the realities of manufacturing the components and assemblies may not allow such superposition due to practical minimum gauge of materials, need for fasteners or special bond layers, increased cross sections to allow for stress concentrations, inability to grow all components to net shape or to shapes that are structurally or functionally optimal, and similar factors. In addition, it is not clear that the costs of manufacturing using these new technologies and materials will show the same weight-cost relation as do current practices.

As a result of the foregoing, while the net effect on cost reduction of new technologies and new materials taken together could well be a factor much greater than 1000, and in principle could approach a factor of 1 million, in practice such large factors may not be attainable. Nonetheless, even "only" a factor of 100–1000 reduction would indeed be a complete revolution.

10 Information Technologies

Time and resource limits for the study made it impossible to address all important areas of technology. The information systems area was not addressed because of the tremendous level of activity in these areas, driven by both commercial and defense sectors. An incomplete list was nonetheless drawn up to indicate where some of the information technology progress could lead and is shown in Table 10.1. One thing is eminently clear: regardless of the demands placed on information storage, processing, and display by the many information-intensive technologies discussed above, the technologies now under investigation and development will be able to meet them. And then some.

Table 10.1. Information Storage, Processing, and Display

- DNA and quantum computers
- Computer interface via voice command and response
- Autonomous, cooperating, space assembly and servicing robots
- Internet-like user-pull information architectures
    - desired result articulated
    - path unspecified
    - method unspecified
- 3-D holographic, life-size, color, full motion displays
- Multiple terabit random access small solid state storage devices
- Biological/silicon-interfaced systems and techniques
- Goal-oriented, self-organizing, learning, adaptable neural nets
- Complete connectivity at >>40 gigabits/second
- Massive bandwidth compression without information loss
- Ubiquitous massively parallel computing
- Many others

Part II

Implications of the New Technologies

11 The Space Industrial Revolution

The aggregate effects of the technologies and developments discussed in Part II, especially the factor of 1000 or more reduction in the cost of building and launching space systems, will trigger a virtual explosion in commercial space activity. New ventures and enterprises—especially those to provide solar power to Earth from space and to provide for mass public space travel—will flourish, and thus the great mass in orbit and the large number and frequency of launches will surpass even the rosiest of today's projections for commercial and government uses of space.

The sheer volume and numbers of tourists enlivened by a new destination will spawn and then stimulate growth in such industries as space hotels and space cruise ships. Space sports pavilions will feature incomparable zero-gravity sporting contests broadcast to TV audiences measured in the billions. Commercially emplaced and operated space business parks and mixed-use manufacturing facilities will be orbited to make products and services for use in space and on the ground. Advertising, film, and television studios will be among the businesses supported by these orbiting business parks. These are in addition to further proliferation of communications, navigation, and position location services using space.

A short study of the effects of these commercial activities done by the author for George Mason University indicates that the mass in orbit will easily grow to be two-to-three orders of magnitude greater than those experienced today as a result of using advanced technologies, if Buckytube materials advances are excluded. (See Fig. 11.1.) If Buckytubes are used to construct most of the hardware of these activities, they reduce by 10–100 times the weight of most spacecraft, phased in over two decades. However, even with such dramatic reductions, the orbited weights will still be 10 to 30 times greater than today, as illustrated in Fig. 11.2. The graph in Fig. 11.3, which shows only the total mass orbited as a function of time, underscores these potential reductions resulting from use of conventional and Buckytube structural materials.

Just as the mass to orbit will increase enormously, so will the number of space launches (always a good measure of activity), which will likewise be measured annually in the tens of thousands compared with less than 100 today. Even when Buckytubes are introduced and used for both launch vehicle and spacecraft construction, the annual launch rates will be in the thousands. (See Fig. 11.4.) While these numbers seem astronomical today, consider the more than 10,000 daily commercial airline flights worldwide today that are simply taken for granted, but were generally unimaginable in the Wright Flyer years. There is no reason why we would not expect space traffic to increase toward comparably high numbers.

This enormously expanded commercial space activity, which will occur in the next several decades, will bring space into the mainstream of everyday experience. In short, space will be used just like other places on the ground and in the air, and will finally become "just another place" rather than the special venue for highly specialized and hugely expensive automated devices, and the

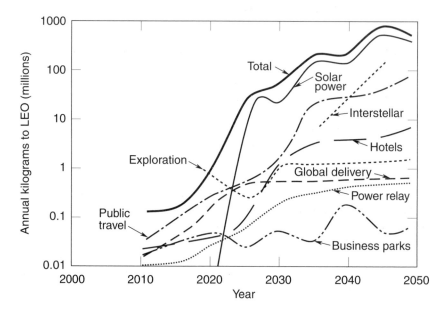

Fig. 11.1. Worldwide potential annual launch weight with conventional structures.

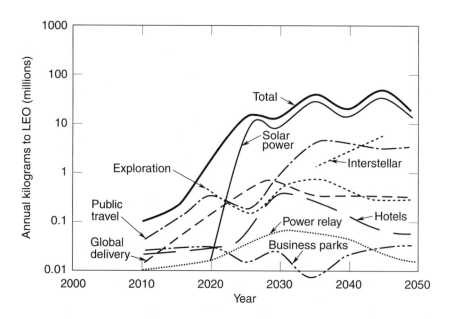

Fig. 11.2. Worldwide potential annual launch weight with Buckytube structures.

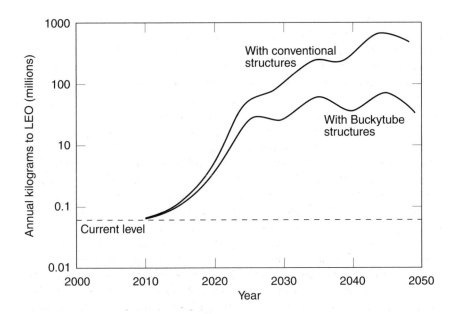

Fig. 11.3. Potential total additive launch weight—all new activities.

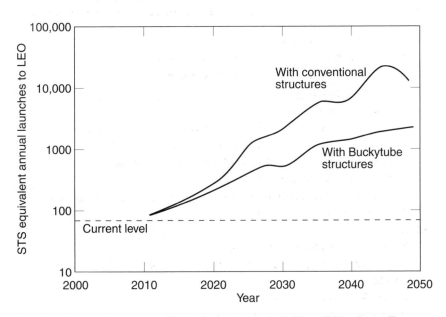

Fig. 11.4. Potential total annual launches—all new activities. (STS—Space Transportation System)

province of a few government astronauts, scientists, and a small number of carefully selected wealthy persons. These activities will undeniably happen because they will be championed by those driven by the entrepreneurial spirit and the quest for profit. The question is not "if" but rather "when," and it is the thesis of this book that they will occur much sooner than anticipated by those steeped in today's technology and programs. They will create unprecedented wealth, solve problems, and bring the benefits of space into the everyday lives of millions of ordinary citizens.

Only then will the public and their representatives be receptive to expanded budgets for exploration, scientific research, and national defense, which are functions traditionally performed by government. Space will be seen as a normal activity of developed nations, much like highways and bridges, for the benefit of everyone and not just for an elite of mainly government astronauts and scientists at great public expense. The excitement of space will return and this time will be available to all. The young of this and other nations see nothing extraordinary with this scenario, and indeed expect it. This space revolution scenario is outlined in Table 11.1.

This "second industrial revolution" will be led by the private sector. It cannot be led by governments, although they would have two critical roles: to lower the technological risks by developing and demonstrating new technologies, and to foster a permissive legal and regulatory environment that enables and supports commercial activities rather than discouraging them.

Table 11.1. The Space Industrial Revolution: The Next 20 to 40 Years

- **Drivers**
  - Entrepreneurial spirit and profit motive
  - Reduced launch costs—up to three orders of magnitude
  - Reduced spacecraft costs—up to two orders of magnitude
  - Blossoming capabilities
  - Commercial uses of space
  - Expansion of personal communications, navigation, observation, data services
  - Public space travel, recreation, hotels, sports pavilions
  - Business parks, advertising, film and television studios, manufacturing
  - Night illumination and energy relay
  - Large-scale generation and transmission of clean energy to Earth power grids
  - Commercial activities on the moon
- **Players**
  - Private sector—leads expansion into space
  - Government—develops technology and fosters business environment
- **Net Benefits**
  - Unprecedented creation of wealth
  - Space in the daily lives of millions of ordinary citizens
  - Proliferating science and defense government programs

12 A New View of Space Emerges

The implications of the foregoing chapters are far-reaching and profound. Use of space will be fundamentally different and quantitatively very much greater in the future, and far beyond our present experience. Future technologies will little resemble linear extrapolations of today's technologies in that space systems and functions will be enormously more capable, affordable, and commonplace. Space will no longer be thought of as extraordinary, exotic, and beyond reach, but will become "just another place" for defense, commerce, research, habitation, and relaxation. Table 12.1 illustrates some of the amazing yet likely potential achievements.

Radically lower costs both to access space and to build and operate space systems will bring about this new space environment. Compared with today's costs, space access for both large and small space systems will appear to be essentially

Table 12.1. Space Features: 2030+

- Adaptive gossamer materials will create antennas 1 kilometer in diameter and telescopes 100 meters in diameter.
- Coherently cooperating picosat swarms will enable huge sparse arrays: RF antennas greater than 100,000 kilometers in diameter and telescopes greater than 100 kilometers in diameter.
- Buckytube materials will reduce the weight of everything by a factor of 100 or more; thus everything will be extremely lightweight and costs will be very low.
- Launch to LEO or GEO will cost no more than $2–20 per kilogram (paralleling airline costs).
- A permanent transportation infrastructure will exist in space—propellantless, reusable, and requiring almost no net energy to operate.
- Many space functions will migrate to GEO and GSO for broad coverage at low cost.
- Launch vehicles will increasingly be electrically and magnetically powered.
- Many sensing and active functions done from aircraft will be shifted to space.
- Spacecraft will be routinely serviced and upgraded, in orbit and on the ground.
- Brassboards and many spares will be routinely orbited.
- Megawatt power levels will be inexpensive and widely available in space.
- Powerful information processing will be available to meet all demands.
- Huge numbers of very small as well as very large lightweight spacecraft will exist.
- Fleets of autonomous robots will fabricate and assemble enormous spacecraft.
- Macrostructures and macroengineering in space will become common.

"free" in the 2030+ time frame. Extremely large and hugely powerful filled aperture space sensing systems will be commonplace because they will be affordable, as well as desirable. In addition, coherent cooperation of many small spacecraft orbited in giant swarms will perform big functions, despite the small size of individual elements. Low cost in conjunction with tripled efficiencies of solar arrays will make available megawatts of power, either directly to individual spacecraft or as beamed power from central space power plants.

The combination of these conditions will enable most functions now performed from LEO or even MEO to be performed equally well, if not better, and at lower expense from GEO. Thus many space functions will migrate to GEO and inclined geosynchronous orbits (GSO), with fewer numbers of larger and more capable spacecraft that are nonetheless far cheaper to build and launch.

In addition, system integration and work hours required to design, develop, and produce space systems will be greatly reduced because spacecraft will be routinely serviced and upgraded in orbit, and relatively unrefined designs and constructions can be tested inexpensively in orbit. The results will be increased numbers of spacecraft, decreased time to upgrade orbital capability, copious use of developmental spacecraft in which space testing is but a routine extension of ground development, and a significant easing of reliability requirements for new space systems. Orbital swarms of tiny spacecraft will make it possible to emplace and replace a few elements or a small portion of the swarm at a time because each functions independently, leaving the function of the entire swarm relatively unaffected during transitions. This capability will also permit incremental acquisition of small portions of the system at a time, reducing the initial costs and spreading the total cost of a system over longer times, thus easing the budgetary approval process.

Perhaps the biggest change will be the presence of an industrial infrastructure in space, which will dominate all defense and civil agency space uses. It will mirror today's ground infrastructure. Tourist, travel, hotel, and power delivery industries will dominate space activities, with their mass and traffic dwarfing those of the defense and civil space programs.

There is a downside of this new view of space and the space industrial revolution, particularly for the military. The global information infrastructure will mature: wideband information will be freely available to all, and the "internet" of the future will instantaneously disseminate details of all developments globally. Thus the technologies and techniques that will make possible these advances will be ubiquitous and cannot be presumed to lend advantage to any one single nation. Further, budget-conscious lawmakers will insist that the enormous capabilities of commercial space systems and infrastructure be used by the military for an ever-increasing array of activities previously reserved for dedicated defense space systems. While the military will be able to gather extensive data and services from these commercial systems, their very existence will seriously hamper the defense establishment's ability to secure approval to develop and operate the specialized systems it may feel it needs. These views are summarized in Table 12.2.

Table 12.2. New View of Space in the Long Term

- Launch is almost free.
- Spacecraft weigh very little for their size.
- Extremely large yet lightweight antennas and optics are commonplace.
- "Big" functions are performed by swarms of cooperating "small" elements.
- Megawatts of lightweight, inexpensive power are available to spacecraft.
- Most functions migrate to GEO, with fewer, but larger spacecraft yet lower cost.
- Brute force, brassboard, and spare spacecraft are commonplace.
- Spacecraft are routinely serviced and upgraded in orbit.
- Spacecraft are emplaced, upgraded, and funded incrementally.
- Commercial energy delivery from and within space is widespread.
- Millions of people travel to space and stay in space hotels as tourists.
- Commercial space infrastructure makes space commerce routine.

BUT

- Wideband information infrastructure is available to anyone everywhere.
- High-resolution images and other space products are available commercially to anyone.
- All nations and groups have access to essentially the same advanced technologies.
- Limited budgets for national security activities force defense agencies to rely heavily on commercial services.

While these capabilities and limitations are somewhat deliberately overstated to make the point, nonetheless they are fairly realistic in the aggregate, and U.S. commercial and defense planners and operators will have to learn to realistically adjust to their general implications.

Part III

Advanced Space Concepts and Applications

13 The Nature of Advanced Concepts

13.1 What Are Advanced Concepts?

This section illustrates how the technologies and techniques described in previous chapters can enable and make practical a number of advanced application concepts to make use of space in the next 30+ years. These concepts were conceived to address objectives relevant to important missions and goals of the defense and civil space agencies and commercial space industries. They are not in any sense comprehensive, nor do they address all or even a major portion of possible goals. Furthermore, while the potential utility or capability of many of these concepts will be evident, this book does not necessarily advocate these concepts nor any particular subset. While the acceptance of any particular concept as attractive and worthy of further attention may well be desired, such acceptance will surely depend on the eye of the beholder. *Thus, these concepts are examples of what could be done if the will and means existed, not necessarily what should be done.*

Creation of advanced concepts followed the guidelines outlined in Table 13.1. They were created to make major changes that shift paradigms, not to make 10-percent or incremental improvements. They should be revolutionary. Incremental improvements in existing systems can be desirable, of course, but they are not the intent of this study. Major changes must be made to produce the new view of space presented in this discussion.

The risk level of advanced concepts and ideas span the range from low to high, as illustrated in Fig. 13.1. While flight programs and even flight demonstrations of technology need to have fairly low risk because substantial funds and near-term operational capabilities hang in the balance, advanced concepts could, and indeed generally should, have a much higher degree of risk. By risk is meant that the probability is high that the concept, as envisioned and analyzed, will either not turn out to be useful or acceptable for adoption into a flight program, or the technology required for its implementation will prove to be too difficult to develop in any given time frame. It is also possible that the jump from technology to system

Table 13.1. What Characterizes Advanced Concepts?

- New capability—paradigm shift
- Order-of-magnitude improvement
- Unconventional—out of the box
- Perhaps high risk
- Perhaps long term
- Activity of any scale—local, national, or international

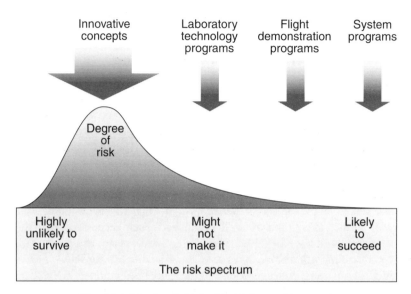

Fig. 13.1. Must seek high risk, high payoff concepts.

implementation will result in a system that, although feasible in principle, may turn out to be so large, heavy, or complex as not to be attractive to implement.

This high risk is precisely what the author intended. There is no shortage of concepts that offer marginal improvements, programs that address the nearer-term and less-risky solutions, and flight programs or political demonstrations of technology that enjoy engineers' confidence. But there are precious few that will take an innovative long-term view and dare to incorporate high risk ideas in the hopes that at least some will prove feasible and reap high rewards. Thus the emphasis of this study—on very ambitious and high risk concepts—is deliberate.

A last context must address the time frame of interest for the concepts. (See Fig. 13.2.) Most space studies address improvements or modifications to current systems, but advanced concepts by their very nature must look much further into the future if they are to live up to the mandate calling for paradigm shifts or order-of-magnitude changes.

In fact, the concepts to be presented span a considerable time range. These time ranges do not assume business as usual. Rather, because advocates will be determined to obtain the benefits of the particular system concept, its technology development, system demonstration, and system development phases will be assumed to be well funded and managed; and once started, the program is left alone, free from interference by requirement changes or politics. Under these conditions, only the difficulty of the technical undertaking will impede the program time line, which will in all cases be considerably shorter than those for programs using current practices. While these are admittedly "best-case" conditions, using them is the only way to normalize the assessment of the time that various concepts will

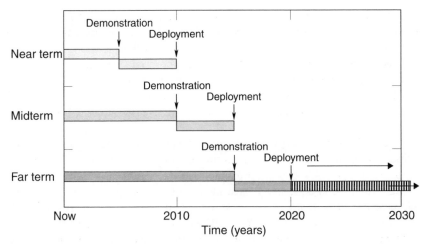

Fig. 13.2. Time frames: programs are assumed to be well funded, well managed, and left alone.

require before they can be fielded. The time-frame labels in the concepts that follow must be viewed in this light.

Some concepts are near term, meaning they could be demonstrated in five years or so and deployed in approximately 10. This implies that the technology must either be in hand or expected to mature within a few years, with five years allocated to system development. (Actually some concepts to be presented do not even require development of new technology or reduction of technology risk. Even though they are known to the technical community, the concepts are not in development for reasons involving politics, agenda, legacy, or simple resistance to different ways of doing things. These concepts are presented within the near-term category, therefore, not because they require technology development, but because they possess substantial unrealized benefits.)

Most concepts presented here do involve a degree of technology development or risk reduction. Such concepts—defined as "midterm"—could be demonstrated in 10 years or so and deployed in approximately 15. Some concepts—defined as "far term"—involve much greater technological difficulty and require 15–20 years until they can be demonstrated and 20–30 years until they could be deployed. In fact some concepts are so ambitious, difficult, or involve such large enterprises that they will probably fall outside of the nominal 30-year horizon of this study. Some of these last concepts are nonetheless included in the "far term" category either because their capabilities are so far reaching or because they represent a wholly new and vital way of perceiving the benefits of space.

Each concept to be presented is labeled as to the likely time frame within which it could be fielded, using the definitions discussed above. Thus the study has a spectrum of concepts with a spectrum of risks, which are applicable to a spectrum of time frames.

Yet another very fundamental aspect of this decision to seek high risk concepts is that many, if not most, of the concepts identified are not expected to make it to flight status. While such low expectation for success is anathema to most success-oriented program managers, it is inherent and appropriate in advanced concepts. If lower risk or a certain percentage of likely successes were required for concept selection, the vitality of the ideas would suffer greatly. Thus if only a few of the many ideas and concepts presented ultimately survive, the study will have been a roaring success.

In fact, this aspect of advanced concepts is well understood in forward-looking entities and is an accepted fact of life in the introduction of new ideas in some industries. Figure 13.3 shows, for example, in the high-tech industry, that on the average, some 3000 conceptual stage ideas will eventually result in one market-able product. Thus, many, if not most, of the concepts to be presented and discussed in the remainder of this book will not make it to the status of operational flight programs. Rather than wringing our hands and bemoaning this fact, we should recognize that this is the expected outcome, and that it is perfectly reasonable. It is important not to let this fact deter concept identification and generation.

13.2 Operative Approaches for Selection of Concepts

Concepts to be discussed were selected from a myriad of possibilities. The guidelines that were generally applied to make these selections are listed in Table 14.2

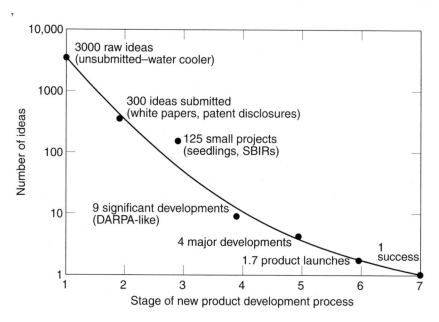

Fig. 13.3. Success curve of innovation.[11.1]

Table 13.2. Operative Guidelines for Identifying Advanced Concepts

- Concentrate on missions that would be revolutionary if feasible.

- Try to do from space whatever can be done from the air or from the ground.

- Perform as many global functions as possible remotely from CONUS (continental United States).

- Aim to reduce or remove crews from harm's way.

- Capitalize on commercial capabilities that are likely to materialize in the far term.

- Revisit previously known but overlooked techniques that are still promising.

- Search for concepts that support unconventional but high-leverage missions such as force projection, as well as the more conventional functions of force multiplication.

- Stay well within the known and accepted laws of physics.

- Pay no attention initially to limitations of policies, politics, or budgets.

- Be "politically incorrect" when necessary. (For example the current aversion to using nuclear reactors in space may be reversed in time as the world geopolitical situation changes, other nations adopt such power, or both.)

with the overarching guideline urging concentration on concepts that would enable revolutionary missions if feasible. With particular attention to defense applications, concepts were identified (or invented) that held promise to accomplish from space whatever could be done from the air or from the ground. Given the radical reductions in cost and increases in capability discussed in the previous chapters, fewer space platforms would be required, and their access to denied areas would be straightforward. Considering the enormous cost of airborne operations, such applications offer a highly promising potential use of space.

Concepts for applications that can perform global functions while being operated in real time from the safety of CONUS were considered highly desirable. Operating space systems from CONUS could remove crews and operators from harm's way by using either uncrewed platforms operated from CONUS or energy or mass delivered directly from CONUS. More general guidelines include capitalizing on commercial capabilities that do not yet exist but are likely to materialize in the long term and revisiting previously overlooked yet still promising techniques whether near or far term.

A more focussed guideline seeks to greatly augment the current use of space—now primarily for applications that serve as force multipliers such as surveillance and communications—with applications that provide direct force such as delivery of energy, weapons, and supplies. While unconventional, some concepts lend themselves best to this use of space and could completely revolutionize military operations. A further guideline bounds concept identification by the known and commonly accepted laws of physics and feasible principles of their application—

use of "Unobtanium" materials or the stuff of science fiction such as postulated "Dilithium crystals" are to be studiously avoided.

It is important to note that however selected, none of these advanced concepts are advocated per se. While many may appear very attractive, the requirements for advocacy are that the system be well understood, its benefits and weaknesses identified and assessed, credible costs for its acquisition developed, and that its place in the strategic plans of the using organization be established, understood, and accepted. Few of the concepts presented can meet most of these criteria. The concepts should thus be taken only as examples of what could be; if they spark interest, they should then be subjected to the definition, trades, design, cost analyses, and assessments that are requisites for advocacy.

Concepts applicable to defense uses must be survivable within their context of use if the nation's defense and safety can be entrusted to them. However, no survivability analyses have been performed for these concepts; therefore, many concepts are presented as though they will undertake to perform the entire military function in a single spacecraft. It could be extremely undesirable to so "place all the eggs in one basket" for reasons of availability as well as survivability. This should not detract from appreciation of the value of many of these concepts, however, because a survivable system can always be defined using the principles of the concept but adding redundancy, proliferation, defensive countermeasures, or other special survivability means or using it only as part of a larger assembly of systems, each of which has different vulnerabilities. The reader must keep in mind that the concepts are simply meant to be examples of new capabilities that could be useful and that might be available with new technologies and innovative implementation.

An additional important, fundamental guideline is to identify concepts without regard for limitations of policies, politics, or even available budgets. There will be no shortage of persons who will apply those criteria after a concept is introduced; therefore, such filters should not be applied by the creators. In fact it is absolutely essential that the concept creation process be free from such limitations or concerns because applying them in advance will inevitably doom many concepts that could actually survive, if not be welcomed with open arms, once the political situation changes. Since such changes are inevitable, it is vital that the concept creation process not be saddled with any current preconceptions of the future climate. In Edward Teller's timeless words "It is extremely hazardous to prognosticate— especially about the future."

13.3 Organization and Context of Advanced Concepts

For ease in correlation with the enabling technologies, the advanced concepts are organized according to the same general plan as the technologies: radio frequency, optics, transportation and infrastructure, and last, energy beaming.

Numbers shown in the illustrations are derived by first-order calculations of the major parameters necessary for feasibility and utility in the application. None are guesses, although detailed calculations such as would be necessary in complete system definition studies were generally not possible, given time and resource constraints.

Weights shown are estimated using the advanced technologies presented in Part II. In all but a few concepts these weights did not include the use of Buckytubes in any fashion; for those that did, that fact was specifically and visibly noted. Thus the generally low concept weights shown are really an upper limit. In order to show the effects of incorporating Buckytubes into the total spacecraft construction, a separate statement immediately follows the weight number, showing the estimated weight if Buckytubes are used. Those weights are generally 100 times lower than those without Buckytube usage. This method was chosen because the weights of some of the large spacecraft will be difficult to accept for many people steeped in today's realities, let alone weights two orders of magnitude smaller yet. Thus, while weights of large spacecraft of only a few kilograms can be fully expected when Buckytubes are used for most of their components, their effect was kept separate in the descriptions to avoid confusion and lack of credibility.

The concepts are shown in a common format, with an illustration for each concept that typically shows its major characteristics, performance, features, and possible application. The caption for each illustration cites a source and the approximate time frame for the concept: near, mid, or far term. The text covers the purpose, benefits, principles, characteristics, performance, weight, technology, and time frame. The illustration, together with accompanying text, thus clearly explains the nature, purpose, and potential value of each concept.

For each application area, the nearest term or simplest concepts are presented first and the longest term, most ambitious, or largest undertakings last. Each concept illustration shows an estimate of the time frame in which implementation might be possible. Although the concepts cover commercial, civilian, and defense uses of space, a majority address defense applications simply because the work for this study was done for The Aerospace Corporation, which operates a federally funded research and development center for DOD. Nonetheless, in the discussion of such concepts, the text identifies those with dual utility for civil or commercial applications, and describes their capabilities.

A number of technologies for concepts that apply to civilian uses of space were originated and/or promulgated by the author while he was Director of Advanced Concepts at NASA Headquarters. Although some of the work on these technologies was not completed when he retired from NASA (January 3, 1997), the author kept in close association with the investigators and applied their results to the concepts in this report.

The author does not claim credit for creating all concepts presented in this book. Some concepts were fully identified by others who were contacted during the research phase of the study; some were conceived by others and subsequently significantly modified by the author—using the technologies and principles addressed in Part II—to make them more practical, more ambitious, or more capable; others were already commonly known and were simply included without changes. Nonetheless, some concepts were indeed conceived by the author based on his experience and understanding civilian and defense space applications, as well as on ideas he had not previously fully formulated or published. Thus the advanced concepts to be presented span a number of sources and different heritages, some of which are identified in each concept discussion.

A last note seems appropriate, based on reaction to these concepts during several dozen presentations given by the author. To borrow (loosely) from Abraham Lincoln:

> All people will love (hate) some of the concepts
> Some people will love (hate) all the concepts
> All people will not love (hate) all the concepts

14 Concepts Using Radio Frequencies and Microwaves

These advanced concepts apply radiation and reception at frequencies varying from ultralow (3–30 hertz) to millimeter waves (30–300 gigahertz). Many capitalize on technologies and techniques described in Part II to make possible higher power, greater apertures, or larger power aperture products than are feasible or have been exploited to date. A number use similar technologies and implementations for the spacecraft, but are applied to such different and unconventional uses that they are shown as separate concepts to illustrate the potential promise of space applications. This is particularly true for a number of the communication concepts.

The 25 concepts in this category are listed in Table 14.1 as near term, midterm, and far term, and are presented in the chapter in that order. They span the range of

Table 14.1. Radio Frequency, Radar, and Communication Concepts

Time Frame	Concept
Near term	Bistatic surveillance using free communications satellites
	Multistatic air vehicle surveillance
	Planetary navigation and communications
	Passive communication satellite using strings of beads
Midterm	Extremely low frequency communications to submarines
	Jam resistant Global Positioning System (GPS) follow-on
	Cockpit weather display
	Unmanned Airborne Vehicle/Unmanned Combat Air Vehicle (UAV/UCAV) control
	Electronic mail transmission
	Foliage penetrating, battle area sensors readout
	Personal communications/wrist radio
	Civil fixed communications/sensor applications
	Civil mobile communications/sensor applications
	Global sonobuoy readout
	Global logistics information system
	Ultra-low frequency radar for submarine detection (U)
	Rotating picosat swarm array RF collector
	High resolution, surface sampling radiometry
	High resolution, surface mapping radiometry
	Rotating nanosat swarm distributed radar
	Bistatic global multifunction radar
	Space traffic and environment control
Far term	High power microwave electronics kill (U)
	Brute force jamming from space
	Active denial fence generator (U)

active and passive RF space devices, with the exception of high power RF transmission and weapon application concepts, which are included in Chapter 17, "Power and Energy Beaming Systems." Detailed descriptions of the few concepts that are classified for national security reasons are obviously not included in this book, but their unclassified titles, indicated by "U" after the title, are listed because they illustrate the scope of the applications.

The concepts are presented approximately in the time frame in which they are likely to be attainable. Within each general time frame category, they are further organized in rough order of increasing complexity, longer time frame, or both. Each concept is described by text and an accompanying graphic illustration.

14.1 Bistatic Surveillance Using Free Communication Satellites

14.1.1 Concept Purpose and Utility

This air and space surveillance capability uses commercial and defense communications satellites as free illumination sources. Its implementation would provide a low cost means to establish a capability to detect small and low observability objects and, in principle, could defeat stealth techniques.

14.1.2 Principles of Operation

To detect very small cross-section objects in air or space, this capability takes advantage of the large enhancement in a RF cross section that occurs when a target is nearly in a straight line between a transmitter and a receiver. This effect of the bistatic cross section enhancement has been known for decades. The enhancement can be 60 decibels (a factor of 1 million) or more, and is a result of the diffraction and forward scattering of the signal around the object. However, the enhancement is limited to angles of only a few degrees from a straight line and so is severely restrictive on the detection geometry. Detection requires that the direct signal be rejected so as not to swamp the much weaker diffracted/scattered signal.

14.1.3 Characteristics of the Illustrated Concept

This space application, illustrated in Fig. 14.1, utilizes the many existing GEO commercial and military communication spacecraft together with those planned to be orbited in LEO (low Earth orbit) as free illuminators. Detection is via antennas and receivers placed on the ground in advantageous fences or patterns. Detection occurs whenever the target, be it orbital debris or air vehicle, passes within a few degrees of a straight line between a satellite and a ground antenna. As there will be many such spacecraft, most on near polar orbits, there will be frequent detection opportunity geometries. Discrimination of the direct signal can be done by pulse modulation and range gating if the transmitter can be controlled, or by using autocorrelation receivers when the transmitter is not controllable.

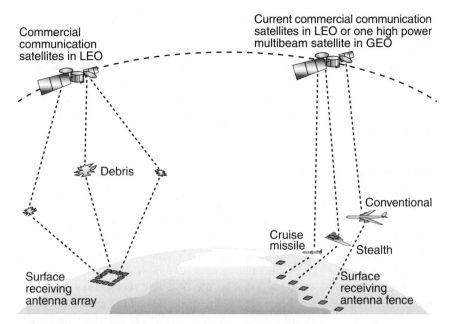

Fig. 14.1. Bistatic surveillance using commercial satcoms.

14.1.4 Performance and Weight

Calculations indicate that cross sections as small as 0.0016 square meters can be detected using 1-square-meter antennas and the radiated effective power densities of currently licensed LEO communication satellites (Iridium, Teledesic, ICO, and others). Future communication satellites at Ku and Ka band could also act as illuminators from MEO (medium Earth orbit) or GEO (geosynchronous orbit), as their delivered power density will be equal if not greater in the future because of the extensive use of many narrow spot beams and large antennas.

This concept could detect orbital debris as small as 4 centimeters across, in time mapping all debris larger than 4 centimeters in orbit. If the detectors and antennas were placed in an appropriate fence, the system could detect all stealth aircraft and cruise missiles with cross sections of 0.0016 square meters or more as they flew between the spacecraft and the ground receivers. The attainment of such small cross sections by air vehicles, particularly as seen from nearly vertical or high angles, is incompatible with also attaining small cross sections in their frontal aspects. Therefore this system can, in principle, detect all low observable and stealth air vehicles.

14.1.5 Technologies and Time Frame

The technologies for this system are in hand. It could certainly be demonstrated by 2005 and deployed by 2010, making this a near term concept.

14.2 Multistatic Air Vehicle Surveillance

14.2.1 Concept Purpose and Utility

Dedicated, space-based, transmitting spacecraft and separate receiving spacecraft or UAV fleets using multistatic radar scatter are employed to detect air vehicle threats. Implementation of this concept would deny the use of stealth and low observables to hostile air vehicles, ranging from large aircraft to small cruise missiles, employed anywhere in the world. It would be a fundamental advantage of great utility to defense.

14.2.2 Principles of Operation

The geometry of these systems is such that the aspect angles at which they view air targets, whether monostatic, bistatic, or multistatic, are usually from above or from the side, avoiding the frontal aspects for which stealth technologies reduce the radar cross section of the air vehicles. This viewing generally "from above" makes the actual physical cross section of the air vehicles determine their radar cross section. The net result is that this concept will enjoy a relatively large radar cross section for most of its targets. The use of a few illuminating spacecraft together with a larger number of receiving spacecraft allows global coverage. UAV fleets can be used in special areas or circumstances to detect the multistatic scattered returns, taking advantage of the shorter range from the targets than would obtain for spacecraft receivers.

While, in principle, bistatic radar could be used, as discussed in the previous concept, the requirement that the target, transmitter, and receiver be within a few degrees of a straight line would constrain geometry to only grazing horizon line of sight paths and might not result in a practical system. The advantage of multistatic over bistatic scatter systems is that they have no such limitations, but are bought at the price of greatly decreased sensitivity due to the lack of the "bistatic cross section enhancement effect." An alternative configuration would place the illuminating transmitters in air vehicles or on the ground and detect the multistatic signals in one or more spacecraft.

14.2.3 Characteristics of the Illustrated Concept

The concept is only notionally presented in Fig. 14.2, as specific system configurations, designs, and detail performance calculations were not performed either for the illustrated concept or for the alternative where the illuminators are in the air or on the ground. Nonetheless this concept is presented in the context of implementing a near term capability using technologies that are mostly available, even though it might have significant performance shortfalls from those ultimately desirable or attainable.

14.2.4 Performance and Weight

Calculations yet to be done.

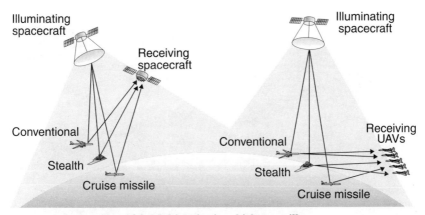

Fig. 14.2. Multistatic air vehicle surveillance.

14.2.5 Technologies and Time Frame

Principal technologies required for the earliest version of this space-based radar concept are mostly in hand. They could probably be demonstrated together by 2005 and deployed by 2010, making this version of the system a near term concept.

14.3 Planetary Navigation and Communications

14.3.1 Concept Purpose and Utility

Very small, low power, small antenna solar system exploration vehicles transmit at high data rates to Earth from planetary distances. Implementation would assure much greater information return from some exploration mission campaigns at much less cost.

14.3.2 Principles of Operation

Many small planetary landers or other exploration vehicles on a planet surface cooperate coherently in transmissions to Earth so that their transmitter powers combine and they form an antenna whose diameter can approach planetary dimensions. This cooperation will allow much more data to be returned to Earth than could be returned by any individual vehicle, and at much reduced total weight and cost. The concept applies to small vehicles in orbit, in the atmosphere, or on the surface of a planet.

14.3.3 Characteristics of the Illustrated Concept

As shown in Fig. 14.3, a small lander transmits its data as a low powered omnidirectional signal, which is received by an orbiting relay/control spacecraft orbiting the planet. That spacecraft transmits control signals to each lander which are retransmitted in an inverse GPS mode so it can determine each lander's exact position. It then commands each lander to control the phase of its transmitter so that the signals from all landers will add in phase in the current direction of the receiving ground station on Earth. When a lander transmits its data, the spacecraft retransmits

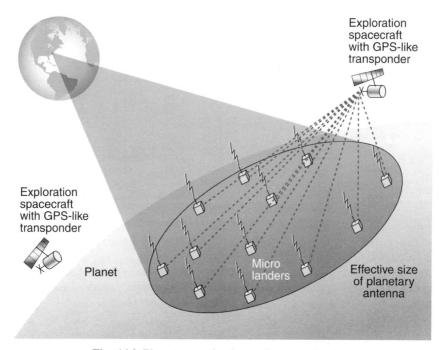

Fig. 14.3. Planetary navigation and communication.

it to all landers, each of which then repeats the data using the proper phase so that all the lander transmissions add coherently at Earth.

What has been achieved is the creation of a coherent sparse array of planetary dimensions, whose beam is so narrow that most of the main beam energy is concentrated on the receiving antenna aperture at Earth. Each planetary orbiter is equipped with GPS-like navigation equipment, forming a planetary navigation grid. A major alternative concept would allow each lander to determine its own required phase shift using the local navigation grid, and then using the orbiting spacecraft only as a repeater the lander sends its signals to Earth through the other landers. The landers and the necessary orbital relay, navigation, and control spacecraft can be emplaced incrementally as missions to the planet are undertaken, so that a planetary capability grows with time.

14.3.4 Performance and Weight

The concept assumes 100 microlanders and microrovers on Mars, each having an omnidirectional antenna, 0.3-watt phase coherent transmitter, transponder, and command receiver. If these were spread over a 500-kilometer-diameter area and transmitted coherently at X band, they would form a 30-watt sparse array whose beamwidth creates a spot at the Earth only 4 kilometers in diameter. Aimed at the Goldstone NASA station the link could support 1 megabit per second data rates from any lander.

14.3.5 Technologies and Time Frame

The technologies for this concept exist but have not been developed and tested as a functioning system. They probably could be demonstrated in space by 2005 and a system deployed by 2010, making this a near term concept.

14.4 Passive Communication Satellite Using Strings of Beads

14.4.1 Concept Purpose and Utility

A completely passive communication satellite, immune to uplink jamming and much harder to destroy with a non-nuclear attack than conventional communication satellites, would result in a harder and more secure complement to high data rate conventional communication satellites.

14.4.2 Principles of Operation

The spacecraft, shown in Fig. 14.4, consists of a number of long tethers attached together and placed in any orbit including GEO. Each tether consists of a nonconducting string nearly aligned with the local vertical to which are attached a large number of small passive reflecting spheres. Each tether thus becomes an end-fire

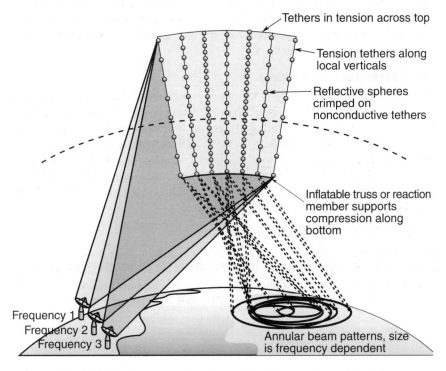

Fig. 14.4. Passive communication satellite using "strings of beads."

reflecting array at RF. When illuminated by a ground transmitter, each array creates an annular beam centered on its axis, resulting in an annulus on the ground.

Different tethers are fitted with spheres of different diameters and at different spacings to form arrays at various operating frequencies. The tethers are stabilized into a passive constellation by a tether connecting their top ends, which tend to spread due to the direction of the gravity gradient at the top. Their bottom ends tend to approach each other and so must be restrained by a compression member, which could be an inflatable boom or virtual boom created by particle exchange dynamics.

14.4.3 Characteristics of the Illustrated Concept

The tether passive antennas could be 100–1000 kilometers long and placed in LEO, MEO, or in GEO. The diameter and annular width of the ground coverage from each of the tether antennas can be varied by changing the frequency of the transmitter. The illuminators could be high power, large aperture transmitters in CONUS, resulting in high power density signals on the ground. The high power annular beam can be placed on a desired receiver for use as a communication system, or an undesired receiver can be positioned in a null.

The "spacecraft" is lightweight and inexpensive, consisting only of tethers and small hollow balls. The entire system is passive and not subject to uplink jamming. Ground receivers are still susceptible to downlink jamming, but the magnitude of the susceptibility may be reduced by the simple expedient of increasing the illuminating transmitter power and/or aperture, something not possible in conventional communication satellites.

14.4.4 Performance and Weight

The performance of this system has not been calculated; however, it was thoroughly analyzed in the 1960s by RAND Corporation when it was seriously proposed as a backbone communication system, but was abandoned when active relay spacecraft became feasible. Nonetheless it may have a limited but important role now to augment high bandwidth communication satellites with a hard, uplink jamming resistant communication relay. A system could weigh 100 to 1000 kilograms. (If implemented using Buckytubes throughout, its total weight would be reduced in the far term to about 1–10 kilograms.)

14.4.5 Technologies and Time Frame

The technologies required are space tethers, which have been proved in 17 successful space flights, and inflatable booms, which have been demonstrated in the NASA Inflatable Antenna Experiment. Thus this system probably could be demonstrated by 2005 and deployed by 2010, making it a near term concept.

14.5 ELF Submarine Communications

14.5.1 Concept Purpose and Utility

This concept presents a means to communicate with submerged submarines any-where in the world from a space platform relay via ELF (extremely low frequency) waves without depending on extremely large above ground surface transmitters; it would also supply a stronger signal to more deeply submerged submarines. Its implementation would complement ground-based so-called "bellringer" systems in disseminating alert and emergency messages to the fleet, and thus increase the robustness of the total capability. The concept is illustrated in Fig. 14.5.

14.5.2 Principles of Operation

A long wire tether in space generates ELF signals, which it can do efficiently since the tether is a conducting wire whose length in space can be made to be a signifi-cant fraction of the transmitted wavelength. These signals propagate along the Earth's magnetic field lines and then bleed through the ionosphere to emerge at its lower boundary. Though the reflection and attenuation of the ionosphere differs greatly from day to night, enough signal exits the lower ionosphere at all times to excite the Earth ionosphere waveguide propagation mode, which has very low

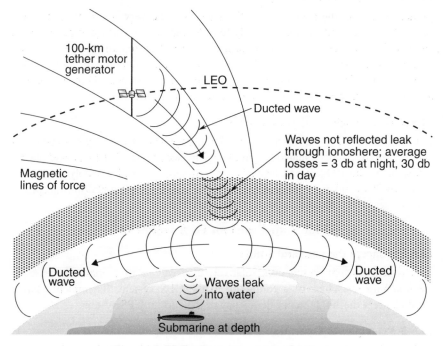

Fig. 14.5. ELF submarine communications.

attenuation and propagates the signals around the world. These signals bleed into the ocean as they propagate, and their low frequency allows them to penetrate the sea water sufficiently to reach submerged submarines.

14.5.3 Characteristics of the Illustrated Concept

The system, shown in Fig. 14.5, utilizes a reversible superconducting or conventional tether to operate in the motor or generator mode. Power at the 100-kilowatt level is generated as a result of magnetic induction when the circuit is closed through the ionosphere via plasma contactors or field emission devices; the tether is thus operated in the generator mode. ELF waves are generated when the tether contact is switched on and off at the ELF rate. The energy both radiated and lost as heat causes the orbit to decay when transmitting. A small solar array powers the tether in the motor mode to raise the orbit and make up the energy lost, but over a longer time after a transmission because the system is expected to operate at a short duty cycle. For an emergency-action message, the spacecraft could be operated continuously until sacrificial reentry. The messages for its transmission originate in ground command transmitters and are simply retransmitted by the spacecraft operating as a relay.

14.5.4 Performance and Weight

Though effective, the bandwidth of the signal will be very small due to the low carrier frequency and can be used only to order the fleet to surface to receive high bandwidth messages through other spacecraft, or to transmit an emergency action message. The system is one way. In these attributes the system concept is no different from today's operational Navy ground-based systems. Only a notional design is shown in Fig. 14.5 to keep the discussion unclassified.

14.5.5 Technologies and Time Frame

Principal technologies for this system are long, large diameter, high power, probably high temperature superconducting wire; long-life electrodynamic tethers; and plasma contactors at the 100-kilowatt level. These technologies could probably be demonstrated by 2010 and a system deployed by 2015, making this a midterm concept.

14.6 Jam-Resistant GPS Follow-on

14.6.1 Concept Purpose and Utility

This system concept presents one relatively easy way to effect major increases in the jamming resistance of a follow-on to the GPS system. Its implementation would correct the single major deficiency of the current GPS system.

14.6.2 Principles of Operation

GPS satellites, shown in Fig. 14.6, are used with a design similar to the current spacecraft but incorporating a large antenna and multiple feeds to create a cellular-like ground coverage over a theater, in which the delivered power density is much

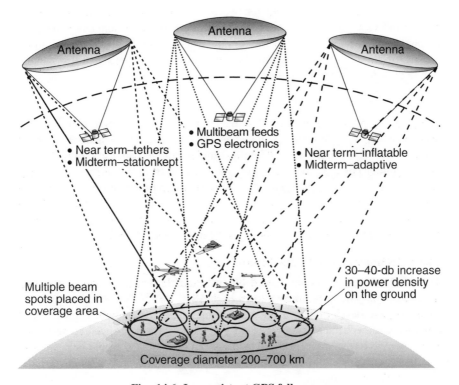

Fig. 14.6. Jam resistant GPS follow-on.

greater than that from the current GPS. Each spacecraft orients its beam pattern to maintain overlapping ground coverage over a theater. The smaller footprint of the larger antenna is replicated at offset angles by the multiple feeds so that a theater substantially larger than the beamwidth is covered.

A transmitter with the same power as today's GPS feeds each beam. The increased power density provided by the concept increases proportionately the jamming power required to jam the downlink. Because the spacecraft antenna is lightweight and inexpensive, the penalties for the larger jamming protection are principally the reduced coverage and the increased total spacecraft transmitted power. The spacecraft are placed into either GEO orbits or into several optional circulating orbits to maintain continuous coverage from the constellations.

14.6.3 Characteristics of the Illustrated Concept

A number of spacecraft constellation configurations are possible. Although the figure illustrates a generic capability, two specific constellations were analyzed. The first places four satellites over a hemisphere—two in GEO orbits 30 degrees apart and two in GSO orbits whose ground track describes a figure 8 centered on the GEO longitude of the node, 180 degrees apart. The second places one satellite in GEO and three in circulating relative orbits whose ground track is a circle around the GEO subsatellite point. In both alternatives the antenna beams are pointed at the theater of interest for continuous coverage. Multiple constellations are deployed, with about six required for illumination coverage such that navigation in theaters worldwide can be supported.

14.6.4 Performance and Weight

The jamming resistance of small terminals in the field is increased by at least 30 decibels when 25- to 35-meter-diameter antennas are used in the spacecraft, and at least 40 decibels or more when 70- to 110-meter antennas are used. In both cases the transmitter power in the spacecraft is unchanged from the current system, but is replicated for every beam generated. This jamming protection is in addition to gains that could materialize using information processing or other means. These gains in performance are large enough that easy and routine jamming of military navigation signals will no longer be a practically attained threat.

14.6.5 Technologies and Time Frame

The 25- to 35-meter antenna could be an inflatable design similar to the IAE (Inflatable Antenna Experiment) flown by NASA. The 70- to 110-meter antenna would require development of an adaptive membrane design and MEMS FEEP (microelectromechanical systems, field emission, electric propulsion) thrusters. Therefore the former concept could be deployed by 2010 and would be near term. The latter probably could be demonstrated by 2010 and deployed by 2015, making it a midterm concept.

14.7 Cockpit Weather Display

14.7.1 Concept Purpose and Utility

Theater and local scale weather is continuously transmitted directly to the cockpits of large numbers of aircraft, where it is displayed as weather maps (see Fig. 14.7). Implementation of this concept would allow the crew to select local or larger scale displays of current weather at their targets, along the expected routes or any other area, resulting in much greater operational effectiveness of military combat and support aircraft. It could, of course, also be extremely useful for civilian government and commercial aircraft operations.

14.7.2 Principles of Operation

A spacecraft in GEO has both a continuous high resolution weather sensor and a wide bandwidth, communication transmission system, so that near real time weather images can be sent to all aircraft in a theater of operations. In use, each pilot would be able to select the weather map of most importance at any time and digitally request it to be sent.

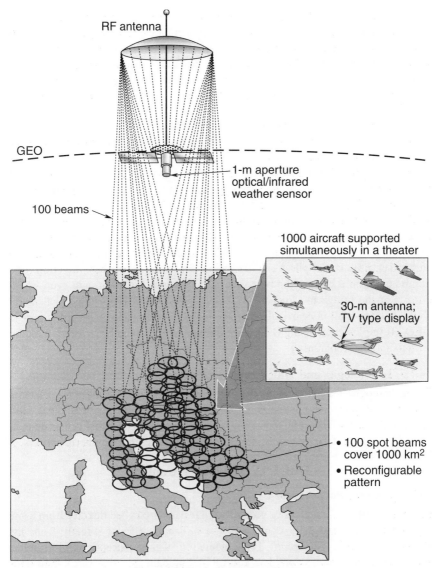

Fig. 14.7. Cockpit weather display.

14.7.3 Characteristics of the Illustrated Concept

A 1-meter aperture optical and infrared weather sensor continuously covers a theater area and produces weather maps with 50-meter resolution, colocated with a communication satellite with a 30-meter-diameter antenna, having 100 separate beams for theater coverage. Every third beam operates on the same frequency so that three frequencies suffice to cover an arbitrarily large area. The antenna could be a space-fed lens or an array-fed reflector, but in either case the large aperture is formed by an adaptive, piezoelectric, electron beam-shaped, membrane technology. The elements of the antenna system can be positioned via a tether against a countermass, and formation flown using MEMS FEEP thrusters, or both, but in neither case is a truss structure needed. The spacecraft transmitters radiate 1 watt each, and the aircraft receivers use 30-centimeter-diameter tracking antennas aboard aircraft platforms.

14.7.4 Performance and Weight

One thousand channels of TV bandwidth images are transmitted simultaneously, and 10 aircraft can be supported simultaneously in each of the 100 separate beams. The total RF bandwidth required for these weather transmissions is only 5 megahertz at S band, due to the very large amount of frequency reuse attained by the antenna. Typical size of a theater covered is 1000 x 1000 kilometers, with 1000 aircraft simultaneously receiving high resolution weather displays updated once every minute. Resolution of the weather information is, at worst, 150 meters in the infrared and 50 meters in the visible region. The net result of the resolution and update time is that pilots can determine, for example, the best direction for a bomb attack on a bridge, the best time and approach direction, and even which portion of the bridge to target for maximum effectiveness, or the equivalent for alternate targets, all in near-real time. The weight is estimated at 700 kilograms in GEO. (If implemented using Buckytubes, total weight would be reduced in the far term to about 7 kilograms.)

14.7.5 Technologies and Time Frame

The technologies used are principally large aperture adaptive, piezoelectric, electron beam-shaped, membrane antennas and MEMS FEEP thrusters, with all other components of essentially current technology. The technologies probably can be demonstrated by 2010 and the system deployed by 2015 making this a midterm concept.

14.8 UAV and UCAV Control

14.8.1 Concept Purpose and Utility

Large fleets of unmanned air or ground vehicles are controlled directly from a site anywhere in a hemisphere containing the utilization theater directly from the CONUS via a relay spacecraft. Implementation of this concept would establish and maintain continuous confident control of, and data reception from, thousands of unmanned vehicles enabling coverage over otherwise denied areas and removing crews from harm's way. The concept is illustrated in Fig. 14.8.

14.8.2 Principles of Operation

The control concept implements a large antenna, high power, communication relay spacecraft stationed in GEO. Its large effective radiated power allows the establishment of high data rate links with many air vehicles, each equipped with modest sized antennas. The high power density projected on the ground comes at

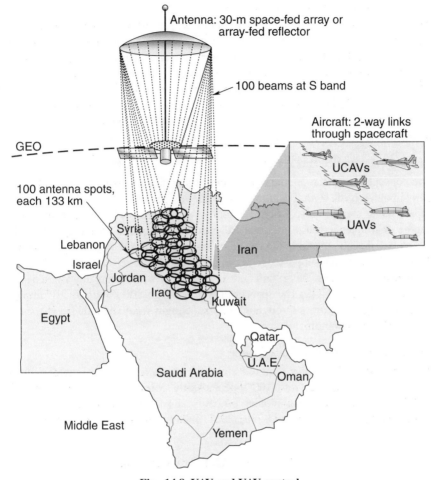

Fig. 14.8. UAV and UAV control.

the expense of a coverage spot much smaller than the size of an average theater, which forces the establishment of many independently steerable beams by the spacecraft antenna. Thus large power density can be distributed in any desired pattern to match the shape of the country or theater being addressed.

The beams can be generated by a fixed feed array illuminating a large reflector, a phased array illuminating the reflector, or by either feed illuminating a lens in a space-fed arrangement. The elements of the antenna system can be positioned via a tether against a countermass, formation flown using MEMS FEEP thrusters, or both. No truss structure is needed. Back links are reciprocal and each feed can have a receiver, with the sum of all receiver outputs sent to the ground where they are separated for individual data stream analysis.

14.8.3 Characteristics of the Illustrated Concept

The conceptual design depicts a spacecraft with a 30-meter-diameter antenna and 100 beams at S band. The size of each antenna spot on the ground is 133 kilometers, and 100 spots can be placed to cover a theater with a maximum extent of 1000 kilometers. Each beam establishes 100 separate channels, each of which is capable of transmitting a 10,000 bits per second data stream, with only 1-watt power per channel. The total power of each beam is 100 watts and the spacecraft total is 10-kilowatt RF. The aircraft requires only a 25-centimeter-diameter tracking antenna and a 1-watt transmitter to close the uplink.

14.8.4 Performance and Weight

The concept can establish continuous control and data readout with 10,000 vehicles simultaneously with a data rate of 10 kilobits per second to and from each, a capacity that should be sufficient for many years to come. Because of the large degree of frequency reuse, the total RF bandwidth required by the spacecraft is less than 1 megahertz, while the total data transferred is 100 megabits per second. This bandwidth can support operation continuously in any theater within the hemisphere containing the ground control station, or with a relay anywhere in the world from the CONUS. The spacecraft would weigh 800 kilograms. (If implemented using Buckytubes throughout, its total weight would be reduced in the far term to about 8 kilograms).

14.8.5 Technologies and Time Frame

The principal technologies required are those of an adaptive, 30-meter-diameter lightweight antenna, with 100 independently pointable beams, and MEMS FEEP thrusters. These technologies probably can be demonstrated well before 2010, and the system deployed before 2015, making this a midterm concept.

14.9 Electronic Mail Transmission

14.9.1 Concept Purpose and Utility

Space and electronic fax media are used to exchange letters between U.S. post offices. Their implementation would greatly speed up mail service without giving up paper as an end medium. The concept illustrated in Fig. 14.9 is an interim system.

14.9.2 Principles of Operation

Fax communications would be established between post offices in cities and larger towns in the United States, interconnecting them through a spacecraft. Paper mail would be automatically read at the source post offices and printed at the destination post offices, with letter carrier delivery. The result would be nearly instantaneous transfer of the "letter" between post offices, with the surface links limited to the drop-off and pickup stations at post offices. This concept could be a stepping stone to the establishment of an all digital, end-to-end, information transfer "post

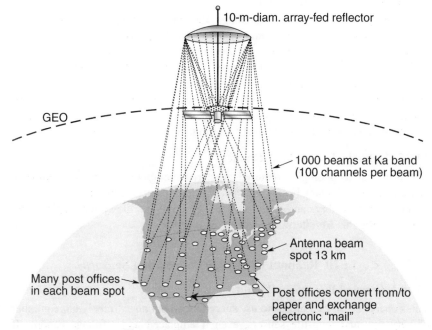

10-m-diam. array-fed reflector

GEO

1000 beams at Ka band
(100 channels per beam)

Antenna beam
spot 13 km

Many post offices
in each beam spot

Post offices convert from/to
paper and exchange
electronic "mail"

Fig. 14.9. Electronic mail transmission.

office." The spacecraft to accomplish the data transfer needs multiple beams, a large antenna, and high power, and would be similar to the spacecraft used in UAV/UCAV control.

14.9.3 Characteristics of the Illustrated Concept

Application of this concept interconnects 100,000 post offices nationwide through one spacecraft in GEO; 1000 separate narrow beams each cover a city—an average city has 100 post offices. Unique coding will enable mail to be routed to the proper destination by an electronic switch in the spacecraft.

The spacecraft has a 10-meter-diameter antenna and 1000 beams at Ka band. The size of each antenna beam spot on the ground is 13 kilometers. Each beam has 100 separate channels, one for each post office in its footprint. Each letter is assumed sent at 10 kilobits per second. Each beam needs to support 100 channels at 0.1 watts using a 1-meter receiving antenna on the ground; thus each beam requires 10 watts and a bandwidth of 1 megahertz. The total RF bandwidth of the system is 3 megahertz, as every third beam can reuse the same frequencies. The total RF power needed in the spacecraft is 10 kilowatts.

The spacecraft has an array-fed reflector antenna of conventional or adaptive design and a very high capacity electronic switch at baseband. Back links are reciprocal and each feed can have a receiver, and the sum of all receiver outputs can be sent to the ground where they are separated for individual switching if desired.

14.9.4 Performance and Weight

This concept attains an exchange among 100,000 U.S. post offices sending and receiving 10,000 letters per second plus 100,000 letter equivalents of junk mail and ads per second. The total system capacity is 30 billion letters and 300 billion junk pieces per year for the United States, attained through a single GEO spacecraft. The post office equipment needed is a transmitter/receiver with 1-meter antenna and 0.1-watt power and two high speed fax machines. The spacecraft would weigh 500 kilograms. (If implemented using Buckytubes throughout, its total weight would be reduced in the far term to about 5 kilograms.)

14.9.5 Technologies and Time Frame

The principal technologies required are large membrane antennas, multibeam phased arrays with many beams, and a very high capacity electronic switch. These technologies probably can be demonstrated by 2010 and the system deployed by 2015, making this a midterm concept.

14.10 Readout of Foliage Penetrating Sensors

14.10.1 Concept Purpose and Utility

This concept presents a means to use space to read out an extremely large number of very tiny and long-life sensors, even if emplaced under foliage canopy. If implemented it could provide highly intrusive and undeniable surveillance of theaters for defense purposes or enable small spatial scale measurements of scientific parameters across a large area of interest.

14.10.2 Principles of Operation

Large numbers of tiny, low power, long-life sensors responding to seismic, acoustic, chemical, or biological signals, or scientific phenomena, would be emplaced from the air or from space and dispersed throughout large areas of interest, including denied areas. In order for the sensor readout function to be reliable, the transmissions would be ultra wideband and in the UHF (ultrahigh frequency) spectrum so as to penetrate foliage. The weak signals would be received by a spacecraft in GEO with a large antenna and relayed to a ground processing site.

14.10.3 Characteristics of the Illustrated Concept

The system conceptual design uses a 300-meter-diameter, adaptive, membrane antenna in GEO, and a number of feeds to generate independent beams. The antenna feed structure is suspended by a tether against a countermass above the antenna, and the elements stabilized and formation flown by MEMS FEEP thrusters. No truss structure is needed.

A conceptual-level design (Fig. 14.10) indicates that the ground sensors need only 30 microwatts of transmitted power into a stub antenna and can operate from any position upon landing under considerable foliage. They each send a 100 bit coded

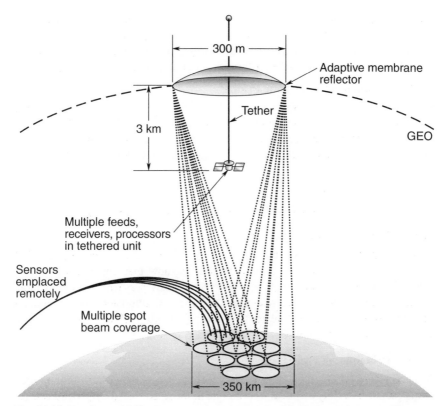

Fig. 14.10. Readout of foliage-penetrating sensors for highly intrusive surveillance in denied areas.

signal, including identification of the sensor, either periodically or when interrogated in sequence by the spacecraft when they detect a signal above their threshold.

The system communication links will sustain 30 million sensors deployed in a 350-kilometer theater, or one sensor every 70 meters on the average. These sensor systems are so small and light that they can even be mounted on cockroaches for continuous dispersal or integrated with biomechanical robots, need only a watch-sized battery, attain a 5-year life even when continuously active, and can be inexpensively produced by the tens of millions for one-time use.

14.10.4 Performance and Weight

The system can read out 30 million sensors every 10 seconds, or 3 million sensors per second. These sensors can be emplaced from aircraft, or to avoid placing crews in harm's way, they could be dispersed by artillery. They could also be emplaced by space plane, or they could be deorbited from spacecraft via a suitably designed reentry vehicle and dispersal mechanism and achieve "highly intrusive surveillance" that cannot be totally denied. Alternatively the system achieves high density monitoring of the atmosphere, weather, oceans, rain forests, eco-logically sensitive preserves, or other science interest areas. The system would weigh 4000 kilograms. (If implemented using Buckytubes throughout, its total weight would be reduced in the far term to about 40 kilograms.)

14.10.5 Technologies and Time Frame

Principal technologies required are adaptive, piezoelectric, electron beam-shaped, very large membrane antenna at UHF, multiple beam feeds, MEMS FEEP thrust-ers, and integrated MEMS sensors and electronics on a chip. These technologies probably can be demonstrated by 2010 and the system deployed by 2015, making this a midterm concept.

14.11 Personal Communications/Wrist Radios

14.11.1 Concept Purpose and Utility

This concept was conceived by the author in 1974 and described in his 1976 "Advanced Concepts" report. It was the first to show that implementation of a "Dick Tracy" wrist radio telephone system for mass personal communications would be feasible using space relays and switches. Iridium later used the concept with low orbit spacecraft and nearer term technology. Its goals have been only par-tially met by cellular telephones and other space systems.

14.11.2 Principles of Operation

A spacecraft in GEO with a large antenna forms many beams whose ground foot-prints establish "cells" identical in principle to those produced by ground cellular systems. These cells can be flexibly placed where needed and will have far fewer dead spots and outages because of the steep propagation angles, compared with ground cellular systems. (See Fig. 14.11.) A large number of cells can be so cre-ated and great frequency reuse attained as a result of the establishment of pre-cisely determined cell boundaries.

Calls received by the spacecraft in one beam and on one frequency are switched onboard and transmitted by the spacecraft on an outgoing beam and frequency to the intended subscriber. Time division or code division multiple access are also feasible. Tens of millions of cellular subscribers can be served by one spacecraft that covers the United States, economically servicing thinly populated as well as dense urban areas. The large antenna allows the subscriber sets to have low RF power and thus become practical to shrink to "wrist radio" size.

14.11.3 Characteristics of the Illustrated Concept

The concept covers the U.S. area with broad beams for rural communications and smaller dedicated beams for cities and towns to handle their larger service demands. This concept incorporates onboard switching to route calls among users and to and from the public switched telephone networks. An alternative leaves the

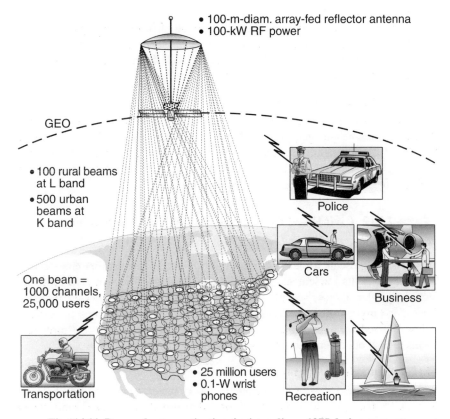

- 100-m-diam. array-fed reflector antenna
- 100-kW RF power

GEO

- 100 rural beams at L band
- 500 urban beams at K band

One beam = 1000 channels, 25,000 users

Police

Cars

Business

Transportation

- 25 million users
- 0.1-W wrist phones

Recreation

Fig. 14.11. Personal communications/wrist radios—1975 design concept.

switching function on the ground at the expense of more bandwidth and time delay to route all calls through the ground switch and back through the spacecraft. The antenna size is sufficiently large that a voice link can be established with, at most, a 0.1-watt transmitter on the user set.

The conceptual design uses a 100-meter-diameter antenna spacecraft in GEO with 100 rural beams covering the United States at L band and 500 urban beams at Ku band placed over the 500 largest cities and towns. The separated portions of the antenna subsystem are supported via a tether and MEMS FEEP thrusters. No truss structure is necessary. The spacecraft has 1,000 100-watt transmitters and a large baseband switch.

14.11.4 Performance and Weight

Each beam supports 1000 voice channels, each of which supports, on the average, 25 subscribers with a 90-percent chance of call completion during the busy hours. The system concept thus supports 25 million subscribers anywhere in the United

States from a single GEO spacecraft. The total bandwidth occupied is only 20 megahertz even though 1800 megahertz are actually being transmitted because of the extensive frequency reuse of the precise beams. The spacecraft would weigh 5000 kilograms. (If implemented using Buckytubes throughout, its total weight would be reduced in the far term to about 50 kilograms.)

14.11.5 Technologies and Time Frame

Principal technologies required are an adaptive, piezoelectric, electron beam-shaped, membrane antenna, many multiple beam feeds or a multibeam feed array, MEMS FEEP thrusters, and very high capacity integrated onboard switches. These technologies probably can be demonstrated by 2010 and the system deployed by 2015, making this a midterm concept.

14.12 Civil Fixed Communication/Sensor Applications

14.12.1 Concept Purpose and Utility

This concept presents a means to read out billions of tiny sensors emplaced in fixed locations. Its implementation would enable highly effective civil functions such as burglar alarms, security fences, automation monitors, border intrusion fences, and myriad other functions that require near continuous, real time monitoring of fixed locations. Implementation of the concept could also provide a means to read out defense fixed sensors.

14.12.2 Principles of Operation

The units are acoustic, seismic, or other sensors integrated in one tiny package with a GPS receiver, a transmitter with omnidirectional antenna, and limited data processing. The use of a large antenna spacecraft receiver enables small transmitted power from the sensor units, which requires only a watch battery. The units will be chip sized and integrated into a package small enough to be disguised as a rock, brick, small plant, or other everyday items, and placed around sites to be protected in multiple fences. They can be indoors or outdoors depending on the security or other function to be performed.

14.12.3 Characteristics of the Illustrated Concept

Each unit is read out by the spacecraft either in sequence, by priority interrogation, or upon a detection message request, at which time the sensor unit transmits its location, identity, and thresholds exceeded in some time period. This data is relayed directly to the ground for analysis and action. The short message lengths mean that a huge number of sensors can be read out, and that large numbers of intrusions can be nearly continuously detected in near real time. (See Fig. 14.12.)

A 100-meter-diameter spacecraft antenna is used, enabling sensor transmitter average power of only 100 microwatts, yielding a 5-year life with a 70-gram battery. The spacecraft antenna generates 200 independent beams which cover the

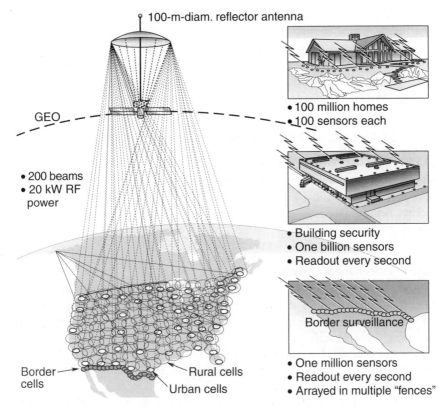

100-m-diam. reflector antenna

GEO

• 200 beams
• 20 kW RF
  power

Border
cells

Rural cells

Urban cells

• 100 million homes
• 100 sensors each

• Building security
• One billion sensors
• Readout every second

Border surveillance

• One million sensors
• Readout every second
• Arrayed in multiple "fences"

Fig. 14.12. Civil fixed communication/sensor applications using tiny, disguised seismic and acoustic sensors with GPS, storage, and communications capability.

entire United States, concentrating with more capacity on the major urban areas and border fences. The spacecraft generates 100 watts of RF power per beam for a total of 20 kilowatts. The sensors are inexpensive "one chip" units powered for 5 years by a watch battery and manufactured by the tens of millions per year. This large production number enables the cost per unit to be drastically reduced to the point that the units are economically expendable. They would be grown in microcircuit furnaces rather than assembled.

14.12.4 Performance and Weight

The illustrated single spacecraft in GEO can read out 10 billion sensors emplaced around 100 million homes, 1 billion urban area sensors around government buildings, and 1 million sensors placed in multiple intrusion fences along a border. Each of these 11 billion sensors can be read out each and every second. The spacecraft would weigh about 5000 kilograms. (If implemented using Buckytubes throughout, its total weight would be reduced in the far term to about 50 kilograms.)

14.12.5 Technologies and Time Frame

Principal technologies required are an adaptive, piezoelectric, electron beam-shaped, membrane antenna, high capacity feed assembly, MEMS FEEP thrusters, high efficiency solar power conversion, and highly integrated onboard electronic switching. These technologies probably can be demonstrated by 2010 and a system deployed by 2015, making this a midterm concept.

14.13 Civil Mobile Communication/Sensor Applications

14.13.1 Concept Purpose and Utility

Billions of tiny microwave tags emplaced in, or attached to, packages, vehicles, and other mobile items would enable the near real time tracking of practically every package being shipped anywhere, vehicles, and transshipment units. Implementation of this concept could also enable the first true speed limit enforcement by allowing localities to remotely set the speed governors of vehicles using their roads.

14.13.2 Principles of Operation

The microwave tags determine their own location using GPS and respond with their identity and location to a spacecraft when it queries them. Each tag is read out by the spacecraft either in sequence or by priority interrogation, at which time the sensor unit transmits its identity code and GPS location. This data is relayed directly to the ground for position determination, tracking, or other functions. The short message time and small message lengths mean that a huge number of tag units can be tracked in real time.

The use of a large antenna spacecraft for reception enables small transmitted power from the tag units, which require only a watch-sized battery. Their signal will be received by the spacecraft while they are in most locations. Metal enclosed warehouses, freight containers, parking garages and the like, which would normally block the transmissions to and from the tag units, could be outfitted with a similar tiny repeater unit in the ceiling or roof with a patch antenna both internally and externally to render the building or container transparent for this function.

14.13.3 Characteristics of the Illustrated Concept

A spacecraft in GEO with a 100-meter-diameter antenna covers the United States, enabling average tag unit transmitter powers of 100 microwatts, yielding a 5-year life with a 70-gram battery. The spacecraft antenna generates 200 independent beams, which cover the entire United States, concentrating on the major urban areas with more capacity. The spacecraft generates 100 watts of RF power per beam for a total of 20 kilowatts, half devoted to urban coverage and half to rural coverage. The concept is similar to the previous concept of fixed civil communications/sensor applications. The sensors are manufactured by the tens of millions per year, drastically reducing costs; they can be economically expended or integrated in products at manufacture. (See Fig. 14.13.)

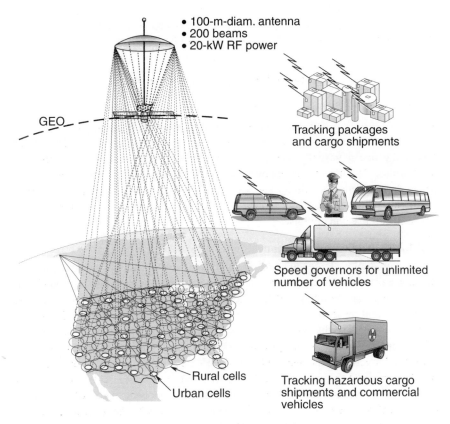

- 100-m-diam. antenna
- 200 beams
- 20-kW RF power

GEO

Tracking packages
and cargo shipments

Speed governors for unlimited
number of vehicles

Rural cells

Urban cells

Tracking hazardous cargo
shipments and commercial
vehicles

Fig. 14.13. Civil mobile communication/sensor applications using tiny, glue-on, reusable "tags" with GPS and communications capability.

14.13.4 Performance and Weight

A single spacecraft in GEO can locate 10 billion tags emplaced in or on 10 billion packages at the rate of 100 million packages every minute, whether in warehouses or in transit, anywhere in the United States. It could also locate 100 million urban area tags and speed governors placed in vehicles for enforcing local speed limits. In addition it could locate 100 million trucks anywhere on or off the roads, at a rate of one per minute. It would weigh about 5,000 kilograms in final orbit. (If implemented using Buckytubes throughout, its total weight would be reduced in the far term to about 50 kilograms.)

14.13.5 Technologies and Time Frame

Principal technologies used in this concept are those of an adaptive, piezoelectric, electron beam-shaped, membrane antenna, high capacity feed assembly, MEMS

FEEP thrusters, high efficiency solar power conversion, and highly integrated onboard processing. These technologies probably can be demonstrated by 2010 and a system deployed by 2015, making this a midterm concept.

14.14 Global Sonobuoy Readout

14.14.1 Concept Purpose and Utility

This concept presents a means to detect, locate, and track essentially continuously all submarines and most large mammals in all oceans. Its implementation would render the ocean virtually transparent for detection and tracking of submerged submarines. Because the sonobuoys can also sense a number of variables such as ocean temperature and salinity and know their position, this capability can result in measurement of winds, ocean currents, and other scientific data. This concept has both defense and civil uses.

14.14.2 Principles of Operation

The system deploys large quantities of small sonobuoys in all the oceans, some free floating and others anchored, that relay acoustic detection data to a ground site through three spacecraft in GEO. The sonobuoys determine their own location using GPS and sense the acoustic environment with passive transducers. They have unique identifiers so that they can be preferentially and unambiguously queried.

The sonobuoy units are integrated packages with a GPS receiver, a transmitter with omnidirectional antenna, and transducers attached to a wire or optical fiber, which is paid out to form an acoustic array. The use of a large antenna spacecraft receiver enables small transmitted power from the sonobuoy units, which require only a watch sized battery for power. Each unit is read out by the spacecraft either in sequence or by priority interrogation, at which time the sonobuoy unit transmits its location, identity, and the results of acoustic detections in some time period. These data are relayed directly to the ground for analysis and action. The short message lengths mean that a huge number of sonobuoys can be read out, and that all submarines and most large marine animals can be tracked continuously.

14.14.3 Characteristics of the Illustrated Concept

A 30-meter-diameter spacecraft antenna is used, enabling average sonobuoy transmitter powers of 2.5 milliwatts, yielding a 5-year life with a 70-gram battery. The spacecraft antenna generates 1000 independent beams covering the oceans in the visible hemisphere; three spacecraft establish a global capability. The sonobuoys are small and lightweight and manufactured by the tens of millions per year, which leads to low production cost making them economically expendable as they are continuously washed up on shore by currents. Continuous replenishment would thus maintain complete coverage as well as fill in coverage voids. They would be emplaced by air drop or deorbited in large clusters from orbiting spacecraft, ballistic vehicles, or space planes. (See Fig. 14.14.)

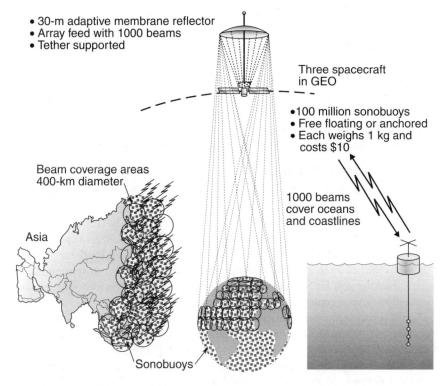

- 30-m adaptive membrane reflector
- Array feed with 1000 beams
- Tether supported

Three spacecraft
in GEO

- 100 million sonobuoys
- Free floating or anchored
- Each weighs 1 kg and
  costs $10

Beam coverage areas
400-km diameter

Asia

1000 beams
cover oceans
and coastlines

Sonobuoys

Fig. 14.14. Global sonobuoy readout for defense and oceanographic applications using inexpensive, expendable sonobuoys.

14.14.4 Performance and Weight

One hundred million sonobuoys can be interrogated every 100 seconds, a rate of 1 million sonobuoys per second. Sonobuoys would be expected to cost no more than $10 each in volume production. The spacecraft weighs 1000 kilograms. (If Buckytubes are used throughout in its implementation, its total weight would be reduced in the far term to about 10 kilograms.)

14.14.5 Technologies and Time Frame

The principal space technologies used in this concept are those of an adaptive piezoelectric, electron beam-shaped, membrane antenna, 1000 beam array or feed assembly, MEMS FEEP thrusters, and high power RF transmission. These technologies probably can be demonstrated by 2010 and a system deployed by 2015, making this a midterm concept.

14.15 Global Logistics Information System

14.15.1 Concept Purpose and Utility

This concept presents a means to read out billions of tiny microwave tags emplaced in, or attached to, packages, vehicles, and other items anywhere on the globe. If implemented the concept would enable the near real time location and tracking of practically every package being shipped anywhere as well as all vehicles, resulting in a true global logistics information system. It could be used by the military, civil government, and commercial entities as common use infrastructure.

14.15.2 Principle of Operation

This system concept is similar to the mobile civil communication/sensor concept described previously, except that a single frequency band is used and the coverage is extended to the entire globe by deploying three spacecraft in GEO. The microwave tags determine their own locations using GPS and respond to a spacecraft's queries with their identities and locations. Each tag is read out by the spacecraft either in sequence or by priority interrogation, at which time the sensor unit transmits its identity and GPS location. These data are relayed directly to the ground for position determination, tracking, or other functions.

The short message time and small message lengths mean that a huge number of tag units can be tracked in real time. The use of a large antenna spacecraft for reception enables small transmitted power from the tag units, which require only a watch sized battery. Their signals will be received by the spacecraft while they are in most locations. Metal enclosed warehouses, freight containers, and parking garages, which would normally block the transmissions to and from the tag units, could be outfitted with a tiny repeater in the ceiling or roof. The concept is illustrated in Fig. 14.15.

14.15.3 Characteristics of the Illustrated Concept

A 100-meter-diameter spacecraft antenna is used on each of three spacecraft in GEO covering the globe with 5000 beams each. The large diameter antenna enables tag unit transmitter powers of 0.4 watts peak but only 100 microwatts average, yielding a 5-year life with a 70-gram battery. Each spacecraft generates a total of 20 kilowatts. Three, or at most four, spacecraft would cover the Earth for global mobile location and tracking services. The sensor tags are inexpensive units that are manufactured by the tens of millions per year. This large production enables the cost to be drastically reduced to the point that the units can be economically expendable or integrated in products at manufacture.

14.15.4 Performance and Weight

The illustrated system can locate 10 billion tags emplaced in or on 10 billion packages at the rate of 100 million packages every minute anywhere in the world, whether they are in warehouses or in transit. The spacecraft would weigh about

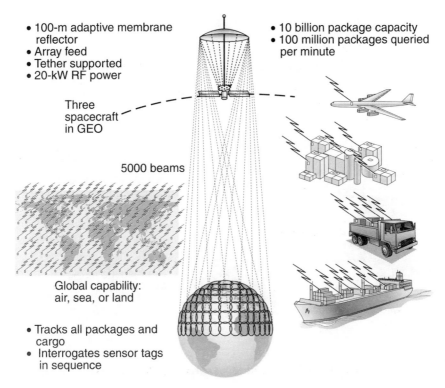

- 100-m adaptive membrane reflector
- Array feed
- Tether supported
- 20-kW RF power

Three spacecraft in GEO

5000 beams

Global capability: air, sea, or land

- Tracks all packages and cargo
- Interrogates sensor tags in sequence

- 10 billion package capacity
- 100 million packages queried per minute

Fig. 14.15. Global logistics information system.

5000 kilograms. (If implemented using Buckytubes throughout, its total weight would be reduced in the far term to about 50 kilograms.)

14.15.5 Technologies and Time Frame

The principal technologies used in this concept are those of an adaptive, piezo-electric, electron beam-shaped, membrane antenna, multiple beam feed array, MEMS FEEP thrusters, high efficiency solar power conversion, and highly integrated onboard processing. These technologies probably can be demonstrated by 2010 and a system deployed by 2015, making this a midterm concept.

14.16 Rotating Picosat Swarm Array Radio Frequency Collector

14.16.1 Concept Purpose and Utility

A truly unconventional, large sparse antenna array RF collector spacecraft with a small surface footprint even when deployed in GEO separates different sources in proximity and also detects weak signals. Its implementation would result in a highly desirable, long dwell RF emitter detection capability. The concept is illustrated in Fig. 14.16.

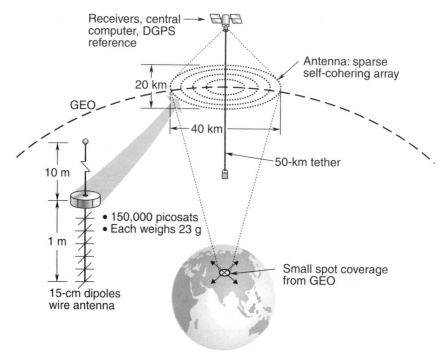

Fig. 14.16. Rotating picosat swarm array RF collector.

14.16.2 Principle of Operation

The large antenna that is the heart of this system is formed by a swarm of tiny elements that make up the lens of a space-fed array with no structure. The antenna is a sparse, self-cohering array formed from a large number of picosatellites rotating (in relative coordinates) in a plane around a central orbital point in GEO. The picosats are self-contained repeater spacecraft. Each receives the ground signal, delays it, and retransmits the signal so that it arrives at the feeds at the same time as a direct ray through the center of the array. The time delay of each picosat is self computed based on its location in the swarm, as measured by a local DGPS (Differential Global Positioning System)-like navigation signal, to compensate for its deviation from its assigned ideal location. Each digitizes, delays, frequency shifts, and retransmits its received signals independently, causing an in-phase composite signal from the ground to be received at the feeds.

The relative positions of these picosat elements change slowly, and only small and infrequent propulsive maneuvers are needed for constellation maintenance. A tether along the local vertical at the central point holds the receivers and DGPS-like reference at the focus against a counterweight. A pseudorandom distribution of the picosats suppresses the antenna grating lobes, and intensive computation greatly reduces much of the remaining sidelobes, creates multiple beams, and steers the ensemble of individual beams anywhere on Earth. The antenna system will function with far fewer elements as a more sparse array, though with limited sensitivity. Thus the system can be incrementally emplaced, upgraded, and even funded with capability growing as budget is available, as opposed to the usual all-or-nothing functioning of today's spacecraft.

14.16.3 Characteristics of the Illustrated Concept

The antenna size is 20 x 40 kilometers and contains 150,000 picosats, each of which weighs 23 grams. The feed array is held in position by a 50-kilometer long, lightweight tether against a counterweight. There is no truss or other structure. Each picosat is gravity-gradient stable, has a dipole array facing Earth, and a broader beam antenna array facing the receivers.

14.16.4 Performance and Weight

The effective collecting aperture of the array is equal to that of an equivalent 80-meter-diameter filled aperture antenna. The coverage spot diameter can be varied by choosing the diameter of the array that is active, with spot sizes on Earth as small as 30 meters at 10 gigahertz, 300 meters at 1 gigahertz, or 3 kilometers at 100 megahertz. It can receive sub-watt signals from individual cell phones. The entire "spacecraft" weighs 3500 kilograms. (If implemented using Bucky-tubes throughout, its total weight would be reduced in the far term to approximately 35 kilograms.)

14.16.5 Technologies and Time Frame

The principal technologies used in this concept are those of coherently cooperating, self-contained, phase or time delay adjusting picosats; long-life tethers to support the electronics and feeds assembly; MEMS FEEP thrusters; highly integrated onboard processing; and integrated single-chip spacecraft. These technologies probably can be demonstrated by 2010 and a system deployed by 2015, making this a midterm concept.

14.17 High Resolution Surface Sampling Radiometry

14.17.1 Concept Purpose and Utility

Highly sensitive radiometry at low microwave frequencies with a small ground footprint would result in high resolution microwave radiometry sampling maps of soil moisture and other surface characteristics, as well as passively detected larger targets.

14.17.2 Principles of Operation

The spacecraft implementation is similar to that of the preceding concept, except that it is designed to map the surface radiation rather than detect discrete emitters. The large antenna, which is the heart of this system, is a swarm of tiny elements forming the lens of a space-fed array with no structure. The antenna is a sparse, self-cohering array formed from a large number of picosatellites rotating (in relative coordinates) in a plane around a central orbital point.

The picosats are self-contained repeater spacecraft. Each receives the ground signal, delays it, and retransmits the signal, causing it to arrive at the feeds at the same time as a direct ray through the center of the array. The time delay of each picosat is self computed based on its location in the swarm, as measured by a local DGPS-like navigation signal, to compensate for its deviation from its assigned ideal location. Each digitizes, delays, frequency shifts, and retransmits its received signals independently, causing an in-phase composite signal from the ground to be received at the feeds.

Only small and infrequent propulsive maneuvers are needed for constellation maintenance. The pseudorandom distribution of picosats suppresses the grating lobes, and intensive computation suppresses much of the remaining sidelobes and steers the beam over the desired swath on Earth. The antenna system will function with fewer elements as a more sparse array with more limited sensitivity but the same directivity. Thus the system can be incrementally emplaced, upgraded, and funded, with capability growing as budget is available, in contrast to the usual all-or-nothing functioning of today's spacecraft. The concept is illustrated in Fig. 14.17.

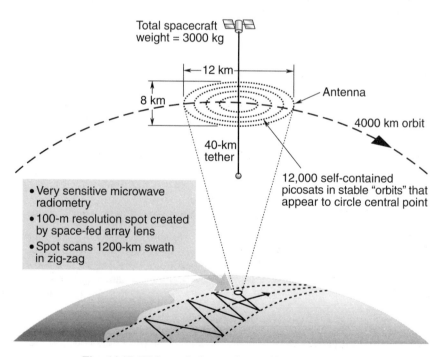

Fig. 14.17. High resolution surface sampling radiometry.

14.17.3 Characteristics of the Illustrated Concept

The antenna size is 8 x 12 kilometers and contains 12,000 picosats, each of which weighs 23 grams. The feed array is held in position by a 40 kilometer long, lightweight tether against a counterweight. The spacecraft scans its coverage spot electronically in a 1200 kilometer zig-zag swath from its 4000 kilometer orbit by modulating the time or frequency shift of the ensemble of picosats. These picosats are similar to those of the preceding concept.

14.17.4 Performance and Weight

The effective collecting aperture of the array is the sum of those of the picosats, and in this example, equal to that of an equivalent 11-meter-diameter antenna. The coverage spot diameter, however, is set by the total aperture diameter of 8 x 12 kilometers, and thus is 100 meters at 1 gigahertz. Five spacecraft would result in a 5-hour global revisit with zig-zag coverage of the scanned swaths. The entire "spacecraft" weighs 3000 kilograms. (If implemented using Buckytubes throughout, its total weight would be reduced in the far term to about 30 kilograms.)

14.17.5 Technologies and Time Frame

Principal technologies in this concept are those of coherently cooperating self-contained, phase or time delay adjusting picosats, long-life tethers to support the electronics and feeds assembly, MEMS FEEP thrusters, and integrated single-chip spacecraft. These technologies probably can be demonstrated by 2010 and a system deployed by 2015, making this a midterm concept.

14.18 High Resolution Surface Mapping Radiometry

14.18.1 Concept Purpose and Utility

Highly sensitive radiometry at low microwave frequencies with a very small footprint on the ground would result in high resolution microwave radiometry maps of soil moisture and other surface characteristics and passively detected larger targets, with 100 percent of Earth's surface mapped with a 5-hour revisit time.

14.18.2 Principles of Operation

The principle of operation is the same as that of the previous concept, except that a multiple element detector array is used in a pushbroom scanning mode for complete Earth coverage rather than only sampling coverage.

14.18.3 Characteristics of the Illustrated Concept

The spacecraft implementation, shown in Fig. 14.18, is similar to that of the preceding concept, except tethers hold a receiving array that must be 2 kilometers long to obtain the 1200 kilometer instantaneous swath width with a resolution of

100 meters. It consists of a 2-kilometer long focal surface with 12,000 printed dipoles, shaped into a focal surface by gravity gradient forces balanced against magnetic forces from a superconducting conductor around its periphery, acting on a piezoelectric, electron beam-shaped, adaptive membrane substrate. The large antenna is formed by a swarm of tiny elements making up the lens of a space-fed array.

The antenna is a 4 x 6-kilometer-diameter, sparse, self-cohering array formed from 12,000 picosatellites weighing 23 grams each, rotating in relative coordinates in a plane around a central orbital point. The picosats are similar to those of the preceding concept. Their locations are initially selected to lie in a plane, and their spacings are pseudorandom to minimize the sidelobe levels, with each picosat designed to loosely stationkeep inside a box 10 meters on a side. The relative positions of these picosat elements changes slowly, and only small and infrequent stationkeeping propulsive maneuvers are needed for constellation maintenance.

14.18.4 Performance and Weight

The effective collecting aperture of the array is the sum of those of the picosats, and in this example, equal to that of an equivalent 6-meter-diameter antenna at 2

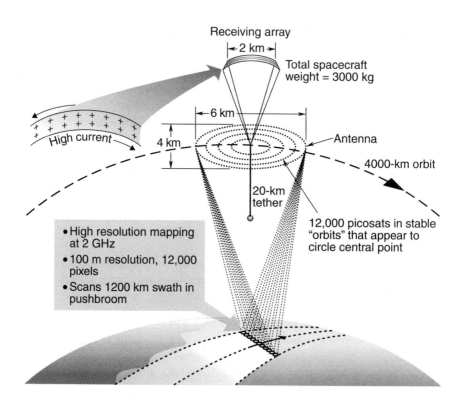

Fig. 14.18. High resolution surface mapping radiometry.

gigahertz. The coverage spot diameter, however, is set by the total aperture diameter of 4 x 6 kilometers, and thus is 100 meters at 2 gigahertz from a 4000-kilometer orbit. Five spacecraft would produce 100-percent global coverage with 5 hour revisit for time critical measurements. The entire "spacecraft" weighs 3000 kilograms. (If implemented using Buckytubes throughout, its total weight would be reduced in the far term to about 30 kilograms.)

14.18.5 Technologies and Time Frame

Principal technologies for this concept are those of coherently cooperating, self-contained, single-chip, phase or time delay adjusting picosats; long-life tethers to support the electronics and feed assembly; MEMS FEEP thrusters; adaptive piezoelectric, electron beam-shaped membranes; and high temperature superconducting coils. These technologies probably can be demonstrated by 2010 and a system deployed by 2015, making this a midterm concept.

14.19 Rotating Nanosat Swarm Distributed Radar

14.19.1 Concept Purpose and Utility

An extremely powerful space-based radar, this concept would allow detection of most air, land, sea, and space targets, as well as many "low observable" targets anywhere, with one or a few spacecraft in GEO.

14.19.2 Principles of Operation

A truly unconventional, large, sparse array antenna using a swarm of nanosats creates a space-based radar system. The spacecraft implementation is similar to that of the preceding rotating swarm concepts, except that it generates and radiates extremely large peak and average powers, and given the generally high angles of viewing can detect and track many air, space, and surface targets from GEO.

14.19.3 Characteristics of the Illustrated Concept

The 10,000 nanosats are self-contained repeater spacecraft weighing about 1 kilogram each. Each spacecraft receives the ground signal, digitizes, delays, and retransmits it, causing it to arrive at the feeds at the same time as a direct ray through the center of the array. The time delay of each nanosat is self computed based on its location in the swarm, as measured by a local DGPS-like navigation signal, to compensate for its deviation from its assigned ideal location. Commands for beam sweep delays are superimposed on the time delays of each nanosat. Each nanosat generates 10 watts of average power and 10 kilowatts peak power at 0.001 duty cycle, has a helical film antenna that increases its gain and doubles as a solar sail for infrequent stationkeeping maneuvers, and has a tether for coarse gravity gradient stabilization. A 50-kilometer tether supports the feed, transmitter, and DGPS reference assembly against a counterweight. The antenna lens is 2 x 4 km. (See Fig. 14.19.)

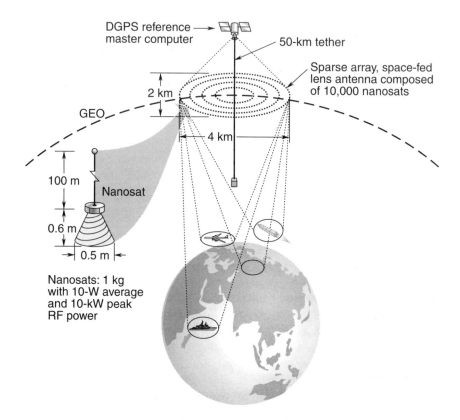

Fig. 14.19. Rotating nanosat swarm distributed radar.

14.19.4 Performance and Weight

The effective area of the array is the same as that of a 50-meter-diameter filled aperture. The total effective RF radiated power of the system is 3 gigawatts peak and 3 megawatts average. Although specific performance calculations have not been performed, these powers are so large that the radar should have the sensitivity from its location in GEO for detecting and tracking many targets simultaneously and most "low observable" targets as well because they are all designed and oriented so as to have their low observables in near-horizontal directions. Three spacecraft would provide essentially complete global coverage. The entire spacecraft weighs about 11,000 kilograms in GEO and can be emplaced and replaced incrementally using small launch vehicles. It could even be funded incrementally. (If implemented using Buckytubes throughout, its total weight would be reduced in the far term to about 110 kilograms.)

14.19.5 Technologies and Time Frame

Principal technologies used in this concept are those of coherently cooperating self-contained, phase or time delay adjusting nanosats, long-life tethers to support the electronics and feeds assembly, MEMS FEEP thrusters, and high power integrated nanosat spacecraft. These technologies probably can be demonstrated by 2010 and a system deployed by 2015, making this a midterm concept.

14.20 Bistatic Global Multifunction Radar

14.20.1 Concept Purpose and Utility

This concept presents a space-based means to replace the functions of all current, planned, and potential AWACS (Airborne Warning and Control System) and JSTARS (Joint Surveillance Target Attack Radar System) aircraft. Its implementation would provide 100-percent global coverage and capability to detect moving targets, as well as perform many radar mapping and space object tracking functions without placing any crews in harm's way.

14.20.2 Principles of Operation

The concept uses three spacecraft with large antennas and high power transmitters in GEO as illuminators and six large collecting aperture, multibeam receiver spacecraft in medium altitude equatorial orbits. (See Fig. 14.20.) It functions in a bistatic or multistatic mode. The very large antennas with their multiple feeds and multiple phase centers and the very high transmitter powers enable the radar to perform both the GMTI (ground moving target indication) functions of the JSTARS aircraft as well as the more difficult AMTI (air moving target identification) functions of the AWACS aircraft. The system performance would enable reliable detection and tracking of reduced cross section and slow moving targets.

14.20.3 Characteristics of the Illustrated Concept

The combination of a 200-meter-diameter active array and 250-kilowatt transmitter results in an effective average radiated power of 7 terawatts. The receiver spacecraft generate multiple independent simultaneous beams while the transmitter spacecraft scan or generate a few beams. Sixty percent efficient, spectrally split, multiple "Rainbow" solar arrays illuminated by a thin film membrane solar concentrator supply the bus power required by the radar transmitters.

Piezoelectric films, adaptively shaped by electron beams, are used to support the active lens elements. These elements correct their phase based on self determination of precise position, using local DGPS-like signals from a self-cohering space-fed lens antenna. Tethers are used to hold the main elements aligned in coarse local vertical, and MEMS FEEP thrusters control attitude and stationkeeping throughout. The

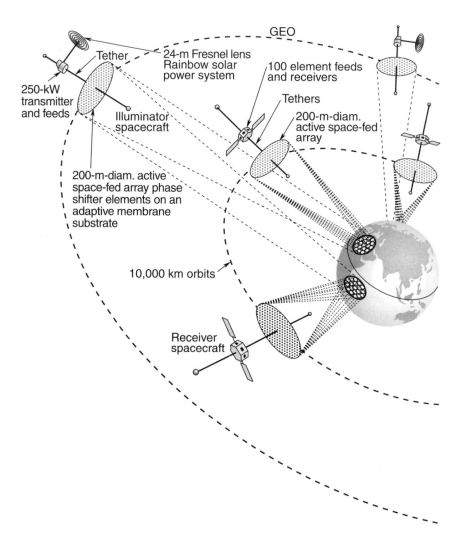

Fig. 14.20. Bistatic global multifunction radar with 7 terawatts of effective radiated power.

transmitter spacecraft weighs less than 10,000 kilograms in GEO, and only three are required. The receiving spacecraft are similar except for generating multiple beams with a phased array feed and lacking the high power supply, and thus lighter.

14.20.4 Performance and Weight

The 7-terawatt effective radiated power and 200-meter-diameter antenna apertures allow the system to duplicate the performance of the airborne AWACS and

JSTARS systems even though the transmitters are at GEO range and the receivers are at 10,000-kilometer altitude. In addition, the high signal levels allow the radar search rates to exceed 1 million square kilometers per second, making the system capable of global utility in several theaters simultaneously. The minimum detectable target velocity and the minimum detectable target size would be reduced by large factors compared with the airborne systems. The ground resolution in the SAR (synthetic aperture radar) mode would be under 10 meters, and complete coverage of the globe to within 20 degrees of the poles is attained.

14.20.5 Technologies and Time Frame

Principal technologies used in this concept are those of large adaptive membrane apertures, MEMS FEEP thrusters, high power and high efficiency rainbow solar power generation, and small active repeater phase, correcting modular lens elements. These technologies probably can be demonstrated by 2010 and a system deployed by 2015, making this a midterm concept.

14.21 Space Traffic and Environment Control

14.21.1 Concept Purpose and Utility

This space traffic and environment control concept presents means for detecting and tracking essentially all space objects, including small debris particles, throughout Earth orbit space. Its implementation would prevent collisions, allow confident space mission planning, detect potentially hostile spacecraft or interceptor maneuvers, detect orbital debris, and control traffic around the International Space Station (ISS) or other high value spacecraft.

14.21.2 Principles of Operation

Radar, infrared, and optical sensor-equipped spacecraft are placed in low polar and equatorial orbits. Search is conducted only above the horizon to minimize ground clutter and background. The spacecraft would be similar to those of the space-based radar system concepts except that far less power and antenna aperture is required as the targets will be viewed against a space, rather than Earth, background.

14.21.3 Characteristics of the Illustrated Concept

The spacecraft were assumed deployed in LEO to obtain Earth limb grazing and higher angle detection against the space background. Other than that, as shown in Fig. 14.21, this system concept is only notionally described—no calculations or sizing were performed. Nonetheless it seems clear that some combination of large diameter, active array, and high power transmitter can result in a high effective radiated power, and thus extreme sensitivity, which in turn leads to detection of small debris particles at all principal altitudes and spacecraft with deliberately reduced cross sections.

Highly efficient, spectrally split, multiple bandgap solar arrays illuminated by thin film membrane solar concentrator power systems would supply the large average

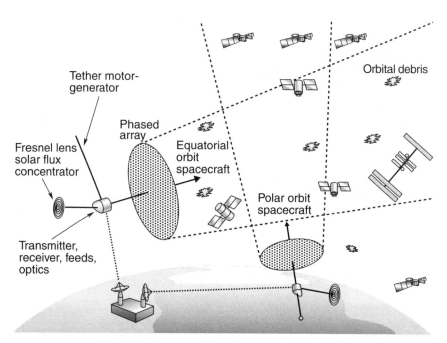

Fig. 14.21. Space traffic and environment control.

average power required by the radar transmitters. Adaptive membrane films would be used to form the antenna substrate, and the antennas could be reflectors or space-fed arrays. Long-life tethers would support all elements, aided by MEMS FEEP thrusters. Optical and infrared sensors would be integrated into the ensemble of structures, either as separate elements or as part of the main assembly. Reversible electrodynamic tether motor-generators would reboost, make up drag, and furnish high peak power to the spacecraft.

14.21.4 Performance and Weight

The combination of optical and infrared sensors, high effective radiated power, and large diameter antenna apertures could allow the system to detect essentially all Earth orbit space objects and debris down to the millimeter size at all altitudes up to GEO. The system would certainly track all spacecraft, including microsats, nanosats, and picosats. Calculations are needed to estimate specific performance and weight of this concept.

14.21.5 Technologies and Time Frame

Principal technologies used in this concept are similar to those of the previous concept: those of large, adaptive membrane apertures, MEMS FEEP thrusters, high power, integrated, high efficiency, rainbow solar power techniques, small integrated repeater, phase correcting, modular lens elements, large phased arrays, and high power, long-life reversible electrodynamic tethers. These technologies could probably be demonstrated by 2010 and a system deployed by 2015, making this a midterm concept.

14.22 Brute Force Jamming from Space

14.22.1 Concept Purpose and Utility

A powerful space-based means of waging electronic warfare with ability to jam most electronics anywhere, this concept would greatly enhance the effectiveness and reduce the cost of tactical operations without exposing crews to harm. The concept is illustrated in Fig. 14.22.

14.22.2 Principles of Operation

The jamming power would be generated by high power transmitters and projected through large steerable beam antennas, and would be powered by high power solar array systems. Operating from MEO, this jammer would have such a large effective

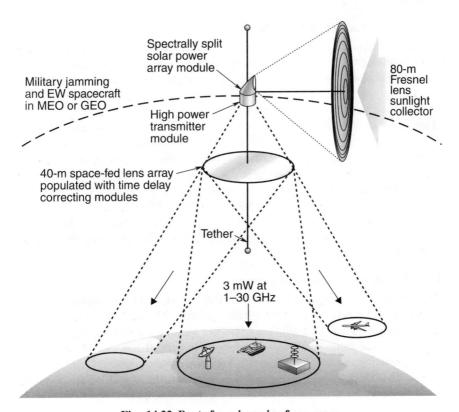

Fig. 14.22. Brute force jamming from space.

radiated power and a small beam footprint that it could be expected to massively overpower the normal received signals, and thus jam and/or spoof most antenna-bearing communication sets, radars, and other electronics on the ground or in the air. The power advantages would be so great that the jammer could usually enter through the sidelobes of the receiving sets' antennas and thus be effective at any incident angle.

14.22.3 Characteristics of the Illustrated Concept

The spacecraft would utilize an 80-meter-diameter, Fresnel lens concentrator, a spectrally split, multiple matched bandgap solar array, and a circulating particle radiator to develop six megawatts of electrical power, which would operate a 3-megawatt RF transmitter and feed array of a space-fed lens antenna. The antenna could be a 40-meter-diameter space-fed lens, constructed of an adaptive piezo-electric electron beam shape-corrected membrane supporting many time delay or phase-correcting electronics modules, and would generate a steerable 3-megawatt beam for EW (electronic warfare) weapon use. This beam would have an effective radiated power of 2400 terawatts at S band.

14.22.4 Performance and Weight

Operating from GEO, these jammers could continuously mainlobe jam and/or spoof essentially all communications and radars in major emplacements, command communications control and intelligence (C3I) centers, field and command locations, ships, and aircraft worldwide with jamming to signal ratios of up to 60 decibels, and without exposing crews to harm. The jamming frequency could be chosen at will by making the transmitter and feed array tunable and the time delay-correcting lens modules broadband. Three such spacecraft would cover the globe for a very effective EW battle support capability. The jamming would be even more effective, jamming through the antenna sidelobes of the ground or air equipment targeted, with jamming-to-signal ratios of 20 decibels or more if the spacecraft altitude were lowered to MEO and five to eight spacecraft orbited for global coverage. The weight of this system would likely be about 9000 kilograms. (If implemented using Buckytubes throughout, its total weight would be reduced in the far term to about 90 kilograms.)

14.22.5 Technologies and Time Frame

The principal technologies used in this concept are those of large adaptive, piezo-electric, electron beam-shaped membrane apertures, MEMS FEEP thrusters; very high power integrated high efficiency rainbow solar power techniques; recirculating particle radiators; small and inexpensive integrated repeater phase or time delay-correcting modular lens elements; high power RF transmitters, and long-life tethers. These technologies could probably be demonstrated by 2015 and a system could be deployed by 2020, making this a far term concept.

15 Concepts Using Optical Wavelengths

These concepts apply transmission and reception at optical wavelengths, generally from LWIR (long wave infrared) to UV (ultraviolet). Many capitalize on the technologies and techniques of Part II to make possible much lighter weight, greater sensitivity, greater apertures, larger intensities, larger fields of view/regard, or larger power than have been realized or exploited to date. The 26 concepts are listed in Table 15.1 according to the estimated time in which they are likely to be attained.

Table 15.1. Optical/Laser Concepts

Time Frame	Concept
Near term	Stereo and interferometric tethered imaging system
	Simple, passive, large constellation
	Water level and fault movement detection
	Covert passive communications
	Night illuminator
Midterm	Satellite to submarine laser communications
	Submarine to satellite laser communications
	Interplanetary TV link
	Simple, large, astronomical space telescope
	Large, simple ground imager
	Simple, distributed, hyperspectral sensor
	Battlefield fear generator
	Laser target designator with remote sensor
	Laser target designator with onboard sensor
	Detection of chemical leaks and releases
	Surface and undersea sensing through clouds
	Minefield and mine detection spacecraft
	Passive aircraft surveillance
	Distributed spacecraft outer planets imager
	Space-based suppression of enemy air defenses (U)
	Large, lightweight, long dwell ground imager
	Large, lightweight, long dwell tethered ground imager
	Large, lightweight astronomical space telescope
	Zooming long dwell ground imager
	Ultraresolution sparse aperture GEO imager
	Ultraresolution ring aperture tethered GEO imager
Far term	Large interferometer to image Earth-size planets around other stars

Within the classifications of near term, midterm, or far term, the concepts are presented in the order of increasing complexity or longer time frame, or both. The concepts in this section span the range of active and passive optical devices using space, except for very high power weapon application concepts, which are included in Chapter 17. A number of the concepts use similar technologies and implementations for spacecraft but are applied to such different and unconventional uses that they are discussed separately to illustrate the potential of space applications.

As in Chapter 14, these titles include a few concepts that are classified for security purposes and whose descriptions are therefore not in this book. Their unclassified titles, so noted by "(U)" after the title, are included in the list to give an indication of the scope of the applications identified. Each concept is illustrated and discussed in the following pages in the order listed in Table 15.1.

15.1 Stereo/Interferometric Tethered Imaging System

15.1.1 Concept Purpose and Utility

Two LEO imaging spacecraft are separated passively by large distances with the lower spacecraft orbiting well inside the sensible drag regions of the atmosphere in a drag-free orbit. Implementation of this concept would obtain higher resolution, real time stereo images than possible with spacecraft in unassisted orbits.

15.1.2 Principles of Operation

Two spacecraft are tethered together along the local vertical, stabilized by gravity gradient forces. The tether is a conducting wire electrodynamic engine that can generate thrust from solar array power without propellants. This thrust overcomes drag from both spacecraft but especially from the lower one, which can then orbit well inside the sensible atmosphere. The two spacecraft together form a stereo imaging system. No propellants are needed to maintain this constellation, and the upper spacecraft acts as the upper tether end mass. The orbit mechanics establish that the center of mass of the system will be in a stable orbit described by a virtual spacecraft with lumped mass equal to that of both spacecraft and the tether.

15.1.3 Characteristics of the Illustrated System

This concept is illustrated in Fig. 15.1. The principal feature is a conducting wire tether to implement an electrodynamic capability in which altitude or inclination can be changed by passing a solar array-generated current through the tether and closing the circuit through the ionosphere. This capability enables the system to overcome drag, gain altitude, or change inclination by interacting with Earth's magnetic field as a linear electric motor without use of propellants. The two spacecraft are attached to the tether in such a manner to allow them freedom to maneuver in attitude. A multistrand Caduceus tether design ensures multidecade lifetime in LEO in the presence of orbital debris.

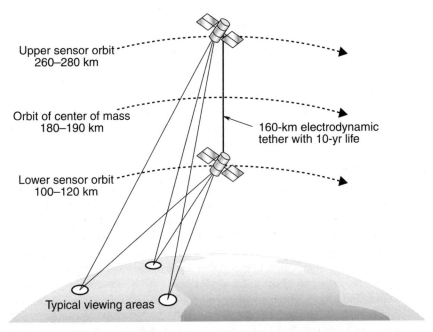

Upper sensor orbit
260–280 km

Orbit of center of mass
180–190 km

160-km electrodynamic
tether with 10-yr life

Lower sensor orbit
100–120 km

Typical viewing areas

Fig. 15.1. Stereo/interferometric tethered imaging system.

15.1.4 Performance and Weight

The lower spacecraft could be orbited at an altitude as low as 100–120 kilometers, limited by aerodynamic heating more than by deceleration forces. A typical configuration has a tether length of 160 kilometers, with the upper spacecraft orbiting at 260 kilometers and the lower spacecraft at 120 kilometers altitude. A tether length of 160 kilometers would result in the tether mass less than 15 percent of that of either spacecraft, using current materials. The tether is extremely inexpensive compared with propulsive means to fly a spacecraft at such low altitudes. The use of a multiline fail safe Hoytether or similar tether design would ensure at least a decade of flight without being cut by orbital debris. (If implemented using Buckytubes throughout, the tether weight would be reduced in the far term to about 0.1 percent of that of the spacecraft.)

15.1.5 Technologies and Time Frame

Principal technologies used in this concept are electrodynamic tethers constructed with multiline fail safe designs. These technologies are to be demonstrated in space by NASA in the near future in the ProSeds (Propulsive Small Expendable Deployer System) program. Currently available materials would be used for the high temperature wire at the tether end. The system probably could be deployed well before 2010 and thus is a near term concept.

15.2 Simple, Passive, Large Constellation

15.2.1 Concept Purpose and Utility

A number of spacecraft are held in a large yet lightweight three dimensional configuration in space without the expenditure of propellants or the use of heavy truss structures to maintain the shape of the constellation. Implementation of this concept would allow the constellations to be used for simultaneous viewing for stereo imaging, interferometric imaging or detection, data fusion, coherent cooperation, or other sensing tasks in which the spacecraft cooperation results in a whole greater than the sum of the parts.

15.2.2 Principles of Operation

The constellation elements are held in place by tension-only tethers, the lightest of all structures. The elements along the local near-vertical are held by gravity gradient forces. Those along the near-local horizontal are held by centrifugal forces due to a slow spin of the constellation. The spin axis is oriented at an angle to the local vertical such that precession maintains the angle constant in an orthogonal direction, stabilizing the entire constellation at that angle to the local vertical as it orbits.

An option replaces two of the horizontal tethers with beams capable of sustaining compression forces. These could be inflatable beams or the virtual beams discussed in Part II, which are formed by reaction forces between opposing circulating streams of particles. With this option the constellation can maintain the three dimensional configuration and arbitrary attitude without rotation.

15.2.3 Characteristics of the Illustrated System

The system is only notionally presented in Fig. 15.2, as no design work was done on a specific illustrative constellation. Any number of spacecraft or sensing or communication elements could be stabilized by these constellation techniques.

15.2.4 Performance and Weight

The overall size of this constellation can be at least 200 kilometers in LEO and 3000 kilometers in GEO without the weight of the tethers exceeding a small fraction of the mass of the spacecraft so supported. These large distances are made possible by the much weaker gravity gradient forces in GEO than in LEO. Even greater separations are possible, but the weight of the tethers, if made of spectra or other conventional materials, will equal the weight of the suspended masses at lengths of 300 kilometers in LEO and 5000 kilometers in GEO, and exceed them as the length increases further. (However, if built out of Buckytube tethers, the constellation could have dimensions on the order of several Earth radii without their weight growing to be a significant fraction of that of the spacecraft.)

15.2.5 Technologies and Time Frame

The tether technology for this concept is either demonstrated already or could be done so in the near future. Neither very long inflatable truss compression beams

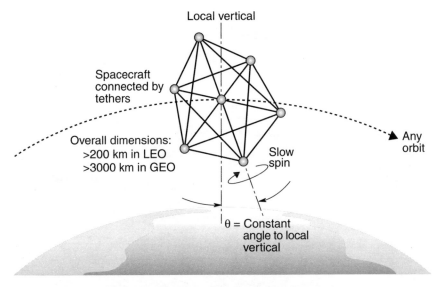

Fig. 15.2. Simple, passive, large interferometer.

nor virtual particle exchange beams have been developed but would be needed only for very largest, nonrotating, three dimensional stable constellations. Otherwise the system probably could be demonstrated in the 2005 time frame and deployed in the 2010 time frame, making it a near term concept.

15.3 Water Level and Fault Movement Detection

15.3.1 Concept Purpose and Utility

This concept presents means for precise and accurate direct determination of water levels in many locations as well as the relative ground movement along many and long earthquake fault lines. The water level information can be used for hydrological maps, flood prediction and control, erosion management, and other aspects of water resources, either alone or acting as ground truth for conventional Earth observing spacecraft. In addition to monitoring crustal activity, earthquake fault movement information could be used in conjunction with other geological data to improve capability to predict earthquakes. The concept is illustrated in Fig. 15.3.

15.3.2 Principles of Operation

The concept employs a picosecond or even a femtosecond pulsed lidar in orbit. The laser pulses are reflected from simple and inexpensive corner reflectors on the surface, and the resulting time delay measured to obtain range (distance) data for each reflector. The location of each reflector is known from emplacement data, and the laser is scanned to illuminate only one reflector at a time. An option exists to passively code the reflections from the reflectors to allow simultaneous interrogation of

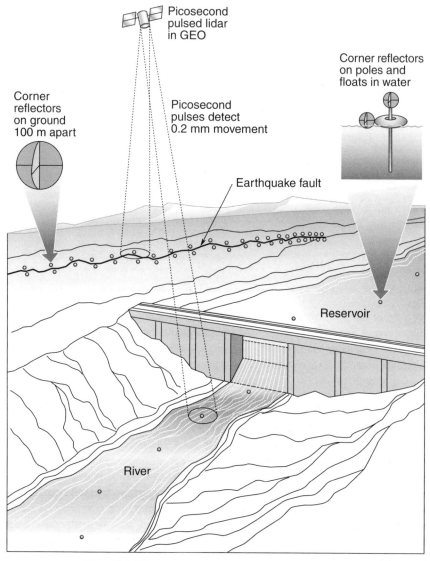

Fig. 15.3. Water level and fault movement detection.

many with a larger beam spot size. Range readings for each reflector are sent to the ground processing center where movements are determined from detected changes.

15.3.3 Characteristics of the Illustrated System

The lidar generates picosecond pulses and detects range differences as small as 200 microns. High incidence angles with respect to vertical would be used for water level measurements. Lower angles would be used for fault line measurements, for which two spacecraft would yield movement data in distance and altitude. The laser can be pointed to scan all instrumented areas sequentially or preferentially by program control, with a pointing accuracy of 2 microradians. Because the laser footprint is 80 meters, the reflectors can be placed 100 meters apart without conflict. The laser power output is nominally 0.1 watts average using 50-centimeter optics diameter, with 0.1 millijoules per pulse occurring at 1 pulse per second.

15.3.4 Performance and Weight

This system has an ability to measure distance to 200,000 reflectors every hour for water level data, where changes can occur in hours in flood situations. Distance measurements on earthquake sensors can be done for a much greater number of reflectors very infrequently because the earthquake sensors move only at a glacial speed, and thus the system can measure 5 million reflectors per day, or 35 million per week. This large capacity, combined with the extreme accuracy of movement detection, will create a powerful new tool for geologic and hydrologic applications not heretofore attainable.

15.3.5 Technologies and Time Frame

The technology for this concept essentially exists, except for the space qualification of laboratory picosecond pulse lidars. This capability probably could be demonstrated in the 2005 time period, and be deployed by 2010, making it a near term concept.

15.4 Covert Passive Communications

15.4.1 Concept Purpose and Utility

Small, lightweight, inexpensive corner reflectors can be used in the field in conjunction with a laser spacecraft to establish a covert, tactical report-back or other communication links with many field elements. Implementation of the concept would result in report-back links that are highly directional, covertly established, jam-proof, and with low probability of being intercepted. (See Fig. 15.4.)

15.4.2 Principles of Operation

The links are established by scanning a laser beam generated on a spacecraft over a tactical or other area of interest. The incident laser light is reflected back to the spacecraft from corner reflectors held by the users. One or more sides of these reflectors are covered with, or manufactured using, arrays of MEMS mirrors whose positions are modulated using voice or digital data signals, or with other means to vary reflectivity. Thus the light reflected back to the spacecraft contains the desired information. The received light is separated from that transmitted at the spacecraft by a filter sensitive only to the modulation and further filtered in the video detector. The system interrogates known reflector user locations or scans given areas at a rate such that the desired information rate is transferred in one or multiple passes.

15.4.3 Characteristics of the Illustrated System

The concept was only notionally designed. While no numbers were calculated, tests of such systems in the laboratory indicate that 10 centimeter, foldable, field reflectors equipped with MEMS mirrors now available commercially could be fielded in the near term. The laser system is unlikely to require more than a watt of average power even when located in GEO, based on similar design numbers of the previous concept. The required pointing accuracy was demonstrated in space by the SDIO RME (Strategic Defense Initiative Organization's Relay Mirror Experiment) program. This system could be used for tracking friendly forces if equipped with the corner reflectors. Others would not be prevented from illuminating the corner reflectors, but coding on the MEMS modulators would ensure acceptance of only permitted units.

15.4.4 Performance and Weight

Performance and weight were not calculated. The system is only notionally defined.

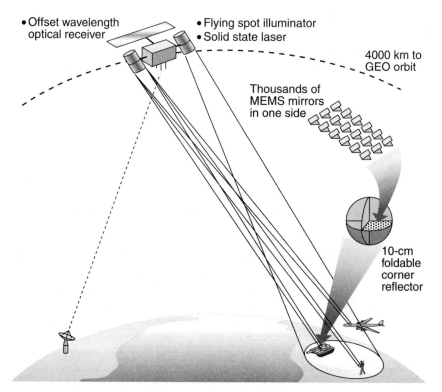

Fig. 15.4. Covert passive communicator.

15.4.5 Technologies and Time Frame

The technologies for this concept essentially exist, except for space qualification of a flying spot laser and heterodyne receiver. The MEMS modulated mirrors are available commercially, though not space qualified. These capabilities probably could be demonstrated in space by 2005 and deployed by 2010, making this a near term concept.

15.5 Night Illuminator

15.5.1 Concept Purpose and Utility

This concept presents means to provide nighttime illumination to large Earth areas without using wires, poles, lights, or energy generation. Its implementation would result in uniform illumination levels over large suburban and rural areas, not just on city streets and primary highways. It would provide for much greater public safety, and save energy and money otherwise required to install, operate, and maintain lighting infrastructure. For military applications, it will deny the cover of darkness and increase security to personnel and installations. The system is illustrated in Fig. 15.5.

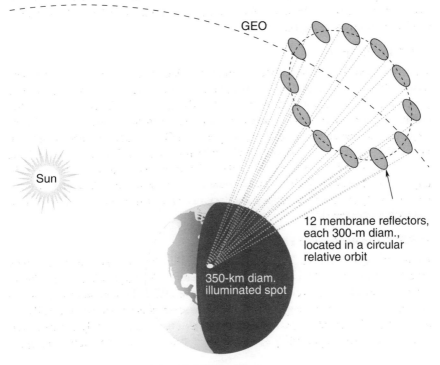

Fig. 15.5. Night illuminator.

15.5.2 Principles of Operation

Sunlight can be reflected toward Earth by an orbital mirror, basically projecting an image of the sun on Earth. The magnitude of this reflected light depends principally on the area of the reflector. The minimum size of the spot created on the ground is set by the 0.5-degree angle that the sun subtends at Earth, therefore creating a beam spread of at least 0.5 degrees from the space craft to the ground. The spacecraft consists of a number of oriented thin film reflectors in GEO, whose light adds in the same spot. The reflectors can be steered so that the spot illuminates any portion of the hemisphere under the spacecraft. The film reflectors can be of poor quality compared with optical flat reflectors without materially increasing the 0.5-degree beam spread and thus the spot size.

15.5.3 Characteristics of the Illustrated System

The system places 12 film membrane reflectors into a circulating relative orbit about a GEO location, in which they require little in the form of maneuvers for stationkeeping. Each mirror is 300 meters in diameter and consists of tensioned flat film held by an inflatable film torus or deployable lightweight truss. The surface accuracy of the reflector film need be no better than about 25 centimeters rms since this is not an imaging system. Each membrane is oriented in attitude using chemical propulsion or ion thrusters, which are also used for the minor translational corrections needed for stationkeeping the reflectors.

15.5.4 Performance

The combined light on the surface from the reflectors is 1 microwatt per square meter, which is a magnitude equal to 10 times that provided by the full moon. It is possible to read a newspaper by this light. The intensity falls off to "only" full moon level for even heavy cloud cover due to scattering from the illuminated cloud layers. The diameter of the light spot from GEO is 350 kilometers and cannot be made smaller by focussing without losing intensity. The illumination level can be varied to simulate a diurnal cycle or for creating effects useful for the military by simply adjusting the angle of some or all the reflectors. A similar concept was seriously proposed to support military operations in Vietnam, although ultimately not developed.

15.5.5 Technologies and Time Frame

All technologies required to field a system for this function exist and have been tested in space, except for the large film reflectors. The large tolerance for figure errors and wrinkles compared with optical or even RF reflectors make this more a packaging than a technology problem. Small scale solar reflectors and sails are beginning to be tested in space. The technologies required probably can be developed by 2005 and deployed by 2010 if desired, making this a near term concept.

15.6 Satellite to Submarine Communications

15.6.1 Concept Purpose and Utility

This concept presents a means to communicate from spacecraft to many submarines while they are submerged at depths of 100 meters or more, with low probability of giving away their locations. Its implementation would enable the establishment of medium rate communication to submerged submarines, in many cases eliminating the need to have the submarines surface to establish such communication data rate, thus removing a major potential submarine operations vulnerability.

15.6.2 Principles of Operation

Modest energy lasers in three spacecraft in GEO point in sequence at known submarine locations as well as at a large number of random spots on the ocean, transmitting at the same encrypted data rate at all times. The illuminated spots would include the known or predicted location of our submarines as well as the many other "empty" positions in order not to give away the location of the submarines. The concept is illustrated in Fig. 15.6.

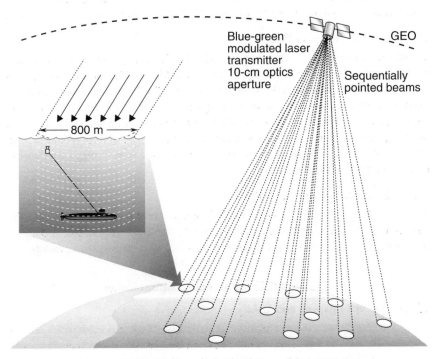

Fig. 15.6. Satellite to submarine laser communications.

A blue-green wavelength laser is used for maximal penetration of sea water, but it limits the use of the system to locations with relatively cloud-free paths to the submarines. An optical sensor viewing upward is placed on the submarine. A detector is also placed on a small platform that can be towed at shallow depths in coastal areas where water turbidity would otherwise severely limit reception at depth. The laser spot can be electronically or electromechanically scanned to point at the various desired locations.

15.6.3 Characteristics of the Illustrated System

The system illustrated in Fig. 15.6 uses a laser aperture of 10 centimeters with an energy of 0.001 joules per pulse and generates 10,000 pulses per second. The average transmitted power is thus 10 watts, while the peak power is 10 kilowatts. The signal is well discernible with a 1-meter-diameter receiving optics with a 1-angstrom bandwidth filter at average depths of 100 meters. Alternatively the receiver could be towed at a 20-meter depth in average murky coastal waters with about the same signal strength received. In the figure, 10,000 spots on the oceans are illuminated, but only 100 of them (1 percent) contain desired submarines. These numbers can be varied in design to be any ratio considered safe, with the rate of access decreasing proportionally with a decrease in the ratio.

15.6.4 Performance and Weight

The illustrated system can send different 25,000-bit messages to each of 100 submarines in 2.78 hours. Alternatively it can send different 250,000-bit messages to each of the 10 submarines in the same time, or to 100 submarines in 16.7 minutes. Sending longer messages to some submarines at the expense of shortening the messages to others is possible but would have to be carefully evaluated because of the possibility of identification of submarine positions. The system would require three spacecraft in GEO for global communications and would relay communications from ground control centers.

15.6.5 Technologies and Time Frame

The technologies for this concept are all in hand except for space qualifying the laser transmitter. This probably could be done by the 2005 time frame, and a system deployed in the 2010 time frame, making this a near term system concept.

15.7 Submarine to Satellite Communications

15.7.1 Concept Purpose and Utility

Lasers used for low rate communications from deeply submerged submarines directly to spacecraft in GEO would allow covert daytime transmissions, which could operate through light cloud cover and fog. Implementation of the concept would allow a covert back link from deeply submerged submarines. This could form the basis for emergency messages or acknowledgement of receipt of higher data rate or priority EAMs (Emergency Action Messages) from command stations without surfacing.

15.7.2 Principles of Operation

The concept, illustrated in Fig. 15.7, uses a blue-green optical wavelength laser in the sail of the submarine and 1-meter optics to direct a beam toward the spacecraft, whose location is known and constant. The beam spreads considerably in passing through the water, and thus pointing can be very crude or perhaps can be dispensed with. The signal arriving in GEO is collected by a large but poor quality photon bucket reflector and passed to a detector through a 1-angstrom filter. The spacecraft points its beam at the desired submarine location and is equipped with a

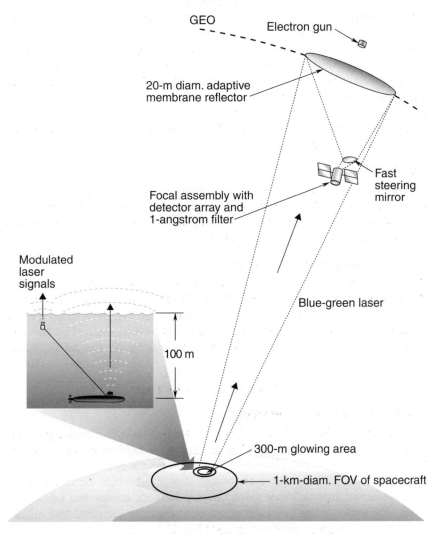

Fig. 15.7. Submarine to satellite laser communications.

fast steering mirror to enable rapid repointing to other submarines. The ocean glows weakly above the transmitting submarine, but its glow is totally masked by reflected sunlight during the day. At night transmissions are interrupted to avoid compromising the submarine's position.

15.7.3 Characteristics of the Illustrated System

The system uses a 1-meter laser transmitter inside the submarine or towed closer to the surface in murky coastal waters. The laser has a 1-joule, pulsed energy output. The spacecraft has a 20-meter-diameter collector/concentrator, which could be a piezoelectric adaptive membrane reflector actuated by an electron beam. The surface quality of this reflector can be 1000 times poorer than one designed for optical imaging because ideally the ground footprint should not be smaller than about 1 kilometer in diameter to simplify pointing and signal acquisition.

A stationkept fast steering mirror using MEMS FEEP thrusters and liquid metal propellants assures that the beam can be pointed rapidly within 500 km and anywhere within an ocean in about 1 minute continuously for many years with little propellant usage. This system attains a link rate of about 2500 bits per second even through moderate clouds and fog. Three such spacecraft in GEO would cover all the oceans on the globe.

15.7.4 Performance

The ability to access one submarine per minute assures that each of 100 submarines will be accessed every 100 minutes during local daylight hours. This system can function as a report-back link working in conjunction with the satellite to submarine forward link of the previous concept.

15.7.5 Technologies and Time Frame

Principal technologies that require development are those of adaptive piezoelectric membranes for large reflectors, MEMS FEEP thrusters for efficient stationkeeping propulsion, 1-joule pulsed lasers for submarine use, and a space qualified 1 angstrom filter. These technologies probably can be demonstrated by 2010 and the system incorporating them deployed by 2015, making this a midterm concept.

15.8 Interplanetary TV Link

15.8.1 Concept Purpose and Utility

A TV bandwidth link available full time would be established from a planetary spacecraft to Earth, using a small, low power laser transmitter on the spacecraft. Its implementation would allow the transmission from a planet of real time, full bandwidth color TV available even through clouds and rain at the receiving station, producing unprecedented video imaging for both science and public relations. The same system can trade signal bandwidth for increased transmission range to cover the entire solar system. (See Fig. 15.8.)

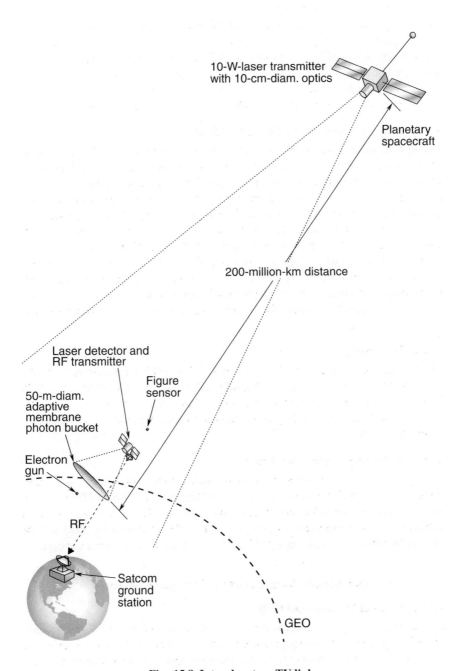

10-W-laser transmitter
with 10-cm-diam. optics

Planetary
spacecraft

200-million-km distance

Laser detector and
RF transmitter

Figure
sensor

50-m-diam.
adaptive
membrane
photon bucket

Electron
gun

RF

Satcom
ground
station

GEO

Fig. 15.8. Interplanetary TV link.

15.8.2 Principles of Operation

The concept envisions a "transformer" in GEO that would translate the optical wavelength signals from the spacecraft at planetary distances into microwave signals that penetrate clouds, resulting in full time availability for the planetary links. Adequate signal strength at the GEO receiving spacecraft for the large bandwidth is attained by using a large but poor optical quality photon bucket. The resulting signals are modulated onto an RF transmitter for all weather transmission to the ground.

15.8.3 Characteristics of the Illustrated System

The system uses a 10-watt laser and 10-centimeter optics on the planetary spacecraft to send information to Earth, with a beam diameter of 1 million kilometers at Earth from a 200-million kilometer distance. Thus the detector in GEO will be able to receive the signal in all parts of its orbit where not eclipsed by the Earth, and the spacecraft beam can be aimed by simply tracking a beacon anywhere on Earth. The use of a 50-meter-diameter photon bucket will allow full 6-megahertz bandwidth TV to be established over a distance of 200 million kilometers.

Because this is a detection rather than imaging system, the 50-meter-diameter collector need only have a surface accuracy of 0.1 millimeter, which is readily attained by a piezoelectric bimorph electron beam-actuated membrane. The detector is a simple photomultiplier tube. All elements are stationkept in formation, and no large truss is used.

15.8.4 Performance

The system could attain a full TV bandwidth at 200 million kilometers, 600-kilohertz bandwidth over a distance of 300 million kilometers, 60 kilohertz over 2 billion kilometers, and 6 kilohertz over 20 billion kilometers with the same collector diameter and laser power and aperture. Data compression algorithms would easily double these distances.

15.8.5 Technologies and Time Frame

Principal technologies that require development are those of adaptive piezoelectric electron beam-shaped membranes for the large reflector, MEMS FEEP thrusters for efficient stationkeeping propulsion, and long-life lasers for planetary application. These technologies probably can be demonstrated by 2010 and the system incorporating them deployed by 2015, making this a midterm concept.

15.9 Simple, Large, Astronomical Space Telescope

15.9.1 Concept Purpose and Utility

A simple, large, lightweight astronomical telescope and scanning spectrometer requiring structural accuracies measured only in centimeters even when sensing in the visible light spectrum would be used in deep space or in solar orbit as a general purpose astronomical telescope.

15.9.2 Principles of Operation

The concept employs a Fresnel zone plate of long focal length as the primary aperture. The plate is a thin membrane coated with aluminized rings that form a Fresnel zone pattern and act as a lens. The focal length of the plate will be very long, and so the focal assembly is stationkept on the optical axis far away, in line with the desired viewing direction. Zone plates are highly dispersive, and so the optical detector will receive a narrow range of wavelengths corresponding to its position on the optical axis.

The optical detector assembly can be translated along the axis resulting in a scanning spectrometer with fine spectral resolution. A Fresnel zone plate has an insertion loss of nearly an order of magnitude so that the primary aperture diameter must be about three times greater than a filled aperture reflective primary for the same sensitivity. Nonetheless, Fresnel zone plates with focal length of thousands of kilometers tolerate relatively large misalignments, and the film surface accuracy and ring location errors can be measured in centimeters rather than in the nanometers typical of astronomical telescopes.

15.9.3 Characteristics of the Illustrated System

The system illustrated in Fig. 15.9 features a 100-meter-diameter film Fresnel zone plate primary optics, with the detector focal assembly stationkept thousands of kilometers away on its optical axis. Scanning of the field of view can be by translation of the focal assembly, with some rotation of the zone plate for large

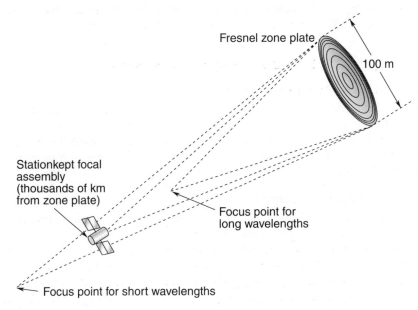

Fig. 15.9. Very simple, large astronomical space telescope.

angle scans. The spectral acceptance band of the detectors could readily extend from the far ultraviolet to the long wave infrared, as a reflecting film sunshade can be interposed between the sun and the zone plate to reduce its temperature and thus its background radiation.

The zone plate film can be flattened to an accuracy of a few centimeters root mean square using an inflated torus or a superconducting coil at its periphery, or the film can be untensioned and composed of a piezoelectric bimorph, actuated by a stationkept electron beam responsive to an optical figure sensor. The zone plate and the focal assembly are oriented and translated using MEMS FEEP liquid metal microthrusters at their periphery.

15.9.4 Performance and Weight

The resolution of the telescope would be 5 nanoradians, while its collecting area would be equivalent to that of a 30-meter-diameter conventional aperture, which is 14 times greater than NASA's planned James Webb Space Telescope. The zone plate will weigh about 400 kilograms, and the total system about 450 kilograms. (If implemented using Buckytubes throughout, its total weight would be reduced in the far term to about 5 kilograms.)

15.9.5 Technologies and Time Frame

The principal new technologies required are those of adaptive piezoelectric electron beam-shaped membranes, MEMS FEEP thrusters, and the materials to allow the membranes to survive for long times in the space environment. These technologies probably can be demonstrated in the 2010 time frame and the system deployed in the 2015 time frame, making this a midterm concept.

15.10 Large, Simple, Ground Imager

15.10.1 Concept Purpose and Utility

A ground imaging spacecraft in GEO utilizes a Fresnel lens to attain larger apertures and lower weights than conventional telescopes. Its implementation would result in a moderate resolution imager for long dwell imaging from GEO, with an on orbit weight lower than that of conventional telescopes, though considerably larger weight than the adaptive film concepts that follow.

15.10.2 Principles of Operation

The concept, shown in Fig. 15.10, uses a Fresnel lens of blazed plastic film as an objective refracting lens, and a relatively conventional telescope as the eyepiece. The film is tensioned by an inflated torus and held by a tether at a distance of several kilometers on the optical axis from the eyepiece telescope. The orientation of the system is roughly along the local vertical at GEO due to gravity gradient forces. The long focal length used results in accuracy requirements for the rings of the Fresnel lens only in the order of several millimeters for imaging at optical wavelengths.

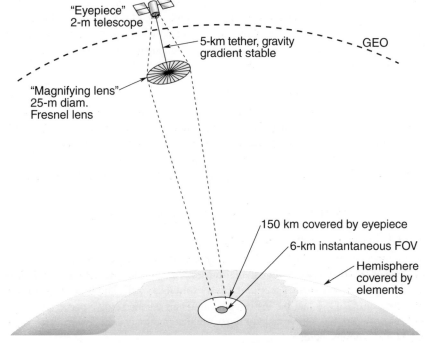

Fig. 15.10. Large, simple ground imager.

15.10.3 Characteristics of the Illustrated System

The Fresnel lens is oriented by MEMS FEEP thrusters at its periphery; it is not spun, and therefore no despin mass is required. The film is flattened by stretching, using an inflated and rigidized torus for support or by a superconducting wire at its periphery. Radial tethers may aid in stiffening the film. Pointing the field of view is accomplished over small displacements by thrusters that move the ensemble relative to the gravity gradient and over large distances by rotating the entire ensemble. These thrusters and their velocity increment are an order of magnitude smaller than if the tether were not used and the two elements simply stationkept relative to each other. The total weight of the spacecraft is on the order of 2000 kilograms using conventional materials.

15.10.4 Performance

The ground resolution of the system is about 2 meters from GEO, with long dwell times attained with a single spacecraft. The system attains an instantaneous field of view of 6 kilometers on the ground. A total field of regard of about 150 kilometers can be attained by moving the eyepiece alone and simply rotating the lens. Much of the visible hemisphere can be covered by slowly rotating the ensemble of elements. Optical performance will be limited by the attainable accuracy of the

Fresnel film plate; however, there will be an irreducible performance degradation caused by scattering from transmissions through the film lens. In addition, sunlight scattering from the film lens can only be avoided by using a sunshade. Though feasible, such a sunshade was not included in the concept described by Lawrence Livermore National Laboratory.

15.10.5 Technologies and Time Frame

Principal technologies required for this concept are those of precision fabrication of the lens membrane and its deployment to maintain the millimeter accuracy including an inflated torus for tensioning, the MEMS FEEP thrusters required for control of the film attitude, and a long-life tether. All these technologies probably can be demonstrated by 2010 and deployed by 2015, making this a midterm concept.

15.11 Simple, Distributed, Hyperspectral Sensor

15.11.1 Concept Purpose and Utility

This concept presents a highly unconventional method of implementing a hyperspectral sensor of great spectral and spatial resolution. Its implementation would allow the detection of very many spectral intervals simultaneously, and it has a small field of view from GEO so that the instrument can dwell on and resolve particular targets of interest. It also has a large field of regard so that one spacecraft covers a significant fraction of a hemisphere.

15.11.2 Principles of Operation

The concept uses a Fresnel zone plate, which is oriented roughly parallel to the local horizontal just below GEO. It is supported by a tether that extends well above the GEO altitude, and may or may not have a counterweight at the top end. The gravity gradient causes the ensemble to remain Earth pointing along the local vertical, with its center of mass in GEO.

The Fresnel zone plate has a long focal length, and thus the surface and ring locations can be imprecise compared with conventional optics. In addition, the lens is a thin film membrane and will be light and inexpensive. It is highly frequency dispersive, and therefore its focal length is a sensitive function of wavelength. Small, self-contained optical sensor nanosats are placed on the tether at many locations, with each nanosat optics filtered for response at only that narrow spectral region focussed at its distance from the lens. The nanosats can transmit directly to the ground, or their signals can be combined in one transceiver, also on the tether.

15.11.3 Characteristics of the Illustrated System

The design illustrated in Fig. 15.11 has a 100-kilometer long tether, which weighs only a few kilograms in GEO. The Fresnel zone plate is 100 meters in diameter, has a collecting aperture equivalent to a 30-meter filled aperture, and requires only a surface accuracy of centimeters in the visible light region. It is constructed of thin film with deposited aluminum rings and is an adaptive piezoelectric membrane

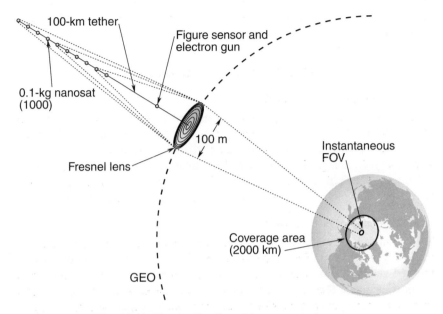

Fig. 15.11. Simple, distributed hyperspectral sensor.

kept flat by an electron beam in response to an optical figure sensor. MEMS FEEP thrusters are at the sensor's periphery for attitude control, with 1000 nanosats attached to the tether. Each nanosat is a self-contained optical sensor spacecraft, integrated into a 0.1-kilogram package. The sensor thus detects 1000 wavelengths simultaneously. The sensor has a resolution as small as 40 centimeters on the ground from GEO. Its field of view can be scanned over an area by tilting the ensemble using MEMS FEEP thrusters on the nanosats, or libration modes can be excited in the tether so that the field of view scans across a 2000-kilometer coverage area in a quasi-random mode, eventually covering the entire area. These thrusters also serve for stationkeeping.

15.11.4 Performance and Weight

The concept can detect 1000 very narrow wavelengths simultaneously with a ground resolution of 40 centimeters and can scan a field of regard of thousands of kilometers. It has the sensitivity given by an equivalent collecting area of a 30-meter diameter in GEO. Its total weight is less than 900 kilograms on orbit. (If implemented using Buckytubes throughout, its total weight would be reduced in the far term to approximately 9 kilograms.)

15.11.5 Technologies and Time Frame

The principal technologies required for this concept are piezoadaptive large membranes actuated by electron beams, MEMS FEEP thrusters, 100-kilometer long, fail safe long-life tethers, and integrated nanosat optical receivers. These technologies probably can be demonstrated by 2010 and the system deployed by 2015, making it a midterm concept.

15.12 Battlefield Fear Generator

15.12.1 Concept Purpose and Utility

A simple and inexpensive psychological, nonlethal weapon with global reach operates by irradiating enemy troops with red or green laser light, and causing some to flee believing that they are being targeted. The effects of this concept have been observed in the field, although the laser light was projected from the ground and from the air rather than from space. Implementation of the concept would extend the effect globally so that it could be exploited anywhere, anytime.

15.12.2 Principles of Operation

An aluminized inflatable adaptive piezoelectric membrane sphere is placed in medium altitude orbit. The shape of at least its lower portion is continuously adjusted via an internal stationkept electron gun in response to an internal figure sensor to maintain accuracy to a fraction of a millimeter. A liquid crystal spatial light modulator layer on its outside surface, responsive to a holographic projector, is used as a second stage of correction so that the reflection phase front is accurate to a fraction of a micron. A ground-based laser of sufficient diameter to produce a spot on the sphere considerably smaller than the sphere's diameter is bounced off the surface. It is aimed at that portion of the sphere that will result in the desired reflection direction, given the orbital path of the satellite.

The reflection from the sphere irradiates the ground at the desired location. The irradiated portion of the sphere can be commanded to be flat for minimal beam spread or curved to create a much larger spot. The result is a bright red spot on Earth trained on troops or civilian concentrations. Based on field experience, many unsophisticated persons seeing that they are illuminated will believe they are being targeted by precision weapons and will often flee in panic. Even if they do not flee the message sent is clear: you cannot hide. If used simultaneously with real weapons in selected locations, the targeted troops never know for sure whether the illumination will precede a lethal attack and cannot take a chance to remain in place.

15.12.3 Characteristics of the Illustrated System

The spacecraft is a 40–meter-diameter inflated thin film balloon, constructed of piezoelectric bimorph materials and adaptively shaped by electron beam in response to an optical figure sensor, both colocated in the sphere's center (Fig. 15.12). A nominal 1-meter-diameter ground laser places a spot no larger than 4 meters in diameter on the sphere, whose local surface curvature is adjusted by command. The liquid crystal spatial modulator layer is illuminated by an optical

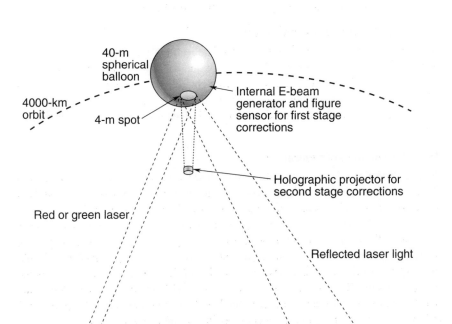

Fig. 15.12. Battlefield fear generator.

projector with a hologram of the receiving errors. Because the spacecraft will be inexpensive, many can be placed in medium altitude orbits for use anywhere on the globe.

15.12.4 Performance and Weight

The spacecraft can place spots on the ground as small as 8 meters in diameter and as large as kilometers. It will weigh less than 300 kilograms. (If implemented using Buckytubes throughout, its total weight would be reduced in the far term to about 3 kilograms.)

15.12.5 Technologies and Time Frame

The principal technologies required are those of piezoadaptive membranes corrected by electron beams, liquid crystals with large correction capability, and MEMS FEEP thrusters for stationkeeping. These technologies probably could be demonstrated by 2010 and deployed by 2015, making this a midterm concept.

15.13 Laser Target Designator with Remote Sensor

15.13.1 Concept Purpose and Utility

A precision tactical target designator, identical in function to that now being used from aircraft, will operate from CONUS (continental United States) or other rearward locations. Implementation of the concept, illustrated in Fig. 15.13, would allow its simultaneous application for overt or covert functions in many areas of the world, including the deepest denied areas. It does not place air crews in harm's way.

15.13.2 Principles of Operation

An adaptive piezoelectric membrane controlled by electron beam in response to a figure sensor and having a liquid crystal front surface layer functioning as second stage corrector is placed in medium altitude orbit. It acts as an optically flat reflector for both the sensing and laser designation functions. An observer or sensor determines the approximate location of a desired target. This area is imaged by using the space system as a mirror, with the imaging sensor on the ground focusing its field of view on the membrane and sensing the portion of Earth reflected by the membrane into its aperture. Because the diameter of the ground field will be exactly twice the diameter of the mirror spacecraft, the ground field will be very small unless the spacecraft is very large (similar to the effect of looking through a soda straw).

The desired area is selected by attitude control of the spacecraft, maintaining the location with continuous adjustment. When the desired target is detected either by a ground operator or by automated trackers, a ground laser beam is reflected off the same space mirror and placed on the desired target by aiming the laser's smaller spot at the proper space location on the space mirror. Closed loop tracking then follows. The pointing and tracking technology was demonstrated by the SDIO (Strategic Defense Initiative Organization) Relay Mirror Experiment.

15.13.3 Characteristics of the Illustrated System

An imaging sensor with 0.6-meter optics sees a 40–meter-diameter field of view on the ground, reflected by a 20–meter-diameter flat film spacecraft. The ground resolution is 1 meter, imaging 1600 x 1600 pixels. When a target is found, a 6–meter-diameter ground laser with 40 watts of power illuminates a 0.5-meter spot on the reflector, in turn placing a 1-meter spot on the target. Closed loop tracking then follows per the normal techniques, with the mirror attitude changed to maintain the laser spot within the mirror diameter as the target moves.

15.13.4 Performance and Weight

A new target can be designated every 2 minutes from the safety of rearward or CONUS areas. One meter spot size is attained and tracking occurs, both without air crews being placed in danger, as missiles or bombs for hitting the target can be released from far away. There are a number of spacecraft in 4000-kilometer orbits for frequent theater coverage. Alternately, one GEO reflector could be used, but it

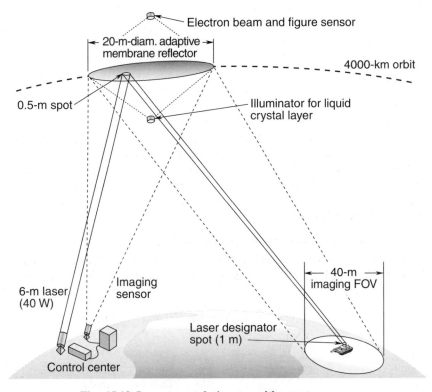

Fig. 15.13. Laser target designator with remote sensor.

would have to be 200 meters in diameter, a much harder task. The total spacecraft weight is 100 kilograms. (If implemented using Buckytubes throughout, its total weight would be reduced in the far term to about 1 kilogram.)

15.13.5 Technologies and Time Frame

The principal technologies required are those of piezoadaptive membranes corrected by electron beams, liquid crystals with large correction capability, and MEMS FEEP thrusters for stationkeeping. These technologies probably could be demonstrated by 2010 and deployed by 2015, making this a midterm concept.

15.14 Laser Target Designator with Onboard Sensor

15.14.1 Concept Purpose and Utility

A precision tactical target designation function identical to that now being done from aircraft could be performed from CONUS or other rearward locations. This concept could be applied simultaneously for overt or covert functions in many areas of the world, including the deepest denied areas. It does not place air crews in harm's way.

15.14.2 Principles of Operation

The concept, shown in Fig. 15.14, is similar to that of the preceding concept except that it greatly broadens its target search sensor coverage by colocating a passive instrument with the reflector spacecraft rather than using one on the ground. An adaptive piezoelectric membrane controlled by electron beam in response to a figure sensor and having a liquid crystal layer for second stage correction, functions as a reflector for the designation laser functions.

An optical, infrared, or multispectral sensor collocated with the spacecraft determines the location of a desired target. When the desired target is detected, the spacecraft mirror attitude is adjusted and the ground laser beam is reflected and placed on the desired target by aiming at the proper spot location on the space mirror spacecraft. Closed loop tracking then follows by a combination of mirror attitude and illumination spot changes. The pointing and tracking technology was demonstrated by the SDIO Relay Mirror Experiment.

15.14.3 Characteristics of the Illustrated System

An imaging sensor with 0.5-meter optics and 10,000 x 10,000 detectors sees a 10–kilometer-diameter field of view on the ground with a 1-meter resolution throughout.

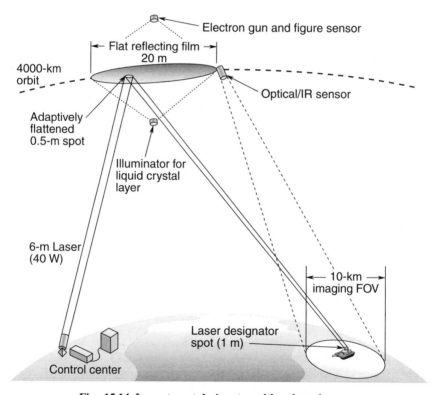

Fig. 15.14. Laser target designator with onboard sensor.

When a target is found, a 6–meter-diameter ground laser with 40 watts of power illuminates an adaptively flattened optical quality 0.5-meter spot on the reflector, in turn placing a 1-meter spot on the target. Closed loop tracking then follows per the normal techniques, with the mirror attitude and spot location changing to maintain the laser spot within the mirror diameter and on the target.

15.14.4 Performance and Weight

A new target can be designated every 2 minutes from the safety of rearward or CONUS areas, within a 10-kilometer theater area. A 1-meter spot size is attained and tracking occurs to an accuracy of 0.5 meter or less. There are a number of spacecraft in 4000-kilometer orbits for frequent theater coverage. The total spacecraft weight is 130 kilograms. (If implemented using Buckytubes throughout, its total weight would be reduced in the far term to about 2 kilograms.)

15.14.5 Technologies and Time Frame

The principal technologies required are those of piezoadaptive membranes corrected by electron beams, liquid crystals with large correction capability, and MEMS FEEP thrusters for stationkeeping. These technologies probably could be demonstrated by 2010 and deployed by 2015, making this a midterm concept.

15.15 Detection of Chemical Leaks and Releases

15.15.1 Concept Purpose and Utility

A space-based hyperspectral sensing system provides detection of spectral signatures of specific chemicals and other aerosol materials when interrogated by laser beams from the ground, tuned to their particular molecular resonant wavelengths. Its implementation would allow performing the same functions as from aircraft, but globally on denied areas and without placing crews at risk. It could also dwell on a suspected target area indefinitely with much better chances of detecting releases into the atmosphere when they occur. It could also detect atmospheric constituents and perhaps some biological agents, and facilitate scientific investigations of atmospheric pollution and other Earth science areas.

15.15.2 Principles of Operation

A large, deliberately highly curved space reflector in GEO illuminates a desired area on the ground, using the space reflector as a means of irradiating the remote ground area from a CONUS-based laser. The curvature of the film is adaptively set and adjusted for reflection of the ground laser to the desired range of ground viewing areas. (See Fig. 15.15.)

The laser spot is reflected from the proper location on the curved reflector to hit the desired ground location. The curvature of the reflector at the laser spot area can be adjusted from flat to curved to change the size of the ground illuminated spot. Since the reflector is in GEO, it can provide continuous access to nearly a hemisphere. The laser is tuned to, or scanned through, the resonant wavelengths,

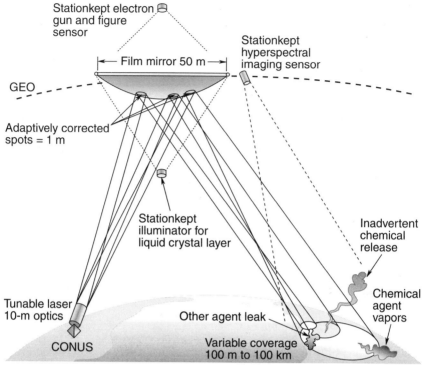

Fig. 15.15. Detection of chemical leaks and releases.

and detects unique molecular signatures that can identify desired effluents or dangerous releases of chemical materials. A hyperspectral imager on or in the vicinity of the spacecraft receives the return from the illuminated regions. It may be possible to extend detection to some bacteriological agents in aerosol form.

15.15.3 Characteristics of the Illustrated System

A 50-meter-diameter, curved piezoelectric membrane spacecraft is used, with adaptive shaping by electron beam in response to an optical figure sensor. The 10-meter-diameter ground laser illuminates a 1-meter spot on the reflector. The curvature of the reflector under this spot is software adjustable by command from flat to curved, and a front surface liquid crystal layer illuminated by a holographic projector functions as a second stage corrector. The stationkept hyperspectral sensor is a self-contained small spacecraft similar to current designs, but uses microfabrication and integration techniques. The laser wavelength is tunable over the range of wavelengths required to detect the phenomena of interest.

15.15.4 Performance and Weight

The ground area illuminated can vary from 100 meters to 100 kilometers in diameter, and can dwell on the target area indefinitely. The laser is tunable to detect molecular resonances. The total weight of the system in GEO is about 400 kilo-grams. (If implemented using Buckytubes throughout, its total weight would be reduced in the far term to about 4 kilograms.)

15.15.5 Technologies and Time Frame

The principal technologies required are those of piezoadaptive membranes cor-rected by electron beams, liquid crystals with large correction capability, and MEMS FEEP thrusters for stationkeeping. These technologies probably could be demonstrated by 2010 and deployed by 2015, making this a midterm concept.

15.16 Surface and Undersea Sensing Through Clouds

15.16.1 Concept Purpose and Utility

A concept using space lasers that penetrate clouds and seawater for surface and undersea detection and imaging functions allows global coverage without the bulk of weather limits and avoids placing air crews in harm's way. The capability would have great utility and could be used in denied areas anywhere at any time.

15.16.2 Principles of Operation

Penetration of moderate cloud thickness by optical wavelengths is possible if ultrashort pulsed lasers are used and the returns from undesirable objects range-gated out. Picosecond or even femtosecond pulse length lasers are now possible, so that range gating allows discrimination of returns separated by substantially less than a meter. With such short pulses, the system can image ground surface features and even detect submerged submarines in the visible light spectrum by rejecting returns scattered from cloud tops and bottoms, the ocean surface, partic-ulates in the clouds and in sea water, and other phenomena. (See Fig. 15.16.)

Implementation of the concept requires a large aperture reflector system for the spaceborne short pulse laser, implemented using a piezoadaptive membrane respon-sive to an electron beam and a figure sensor, and liquid crystal spatial light modula-tor at the laser to predistort its output to correct for the remaining membrane surface errors. MEMS FEEP thrusters are used to position the membrane in GEO.

15.16.3 Characteristics of the Illustrated System

A 10-meter-diameter adaptive reflector is used in the spacecraft. A stationkept fast steering mirror is used to steer the instantaneous field of view to the desired ground location or scan the desired theater area. The elements are stationkept with respect to each other, and focal length is continuously adjustable to effect the spot size of choice.

15.16.4 Performance and Weight

A picosecond or femtosecond pulsed laser stationkept beneath a 10-meter adap-tive membrane reflector optics should penetrate kilometer-thick clouds as well as sea water to significant depths, limited mostly by available pulse power density.

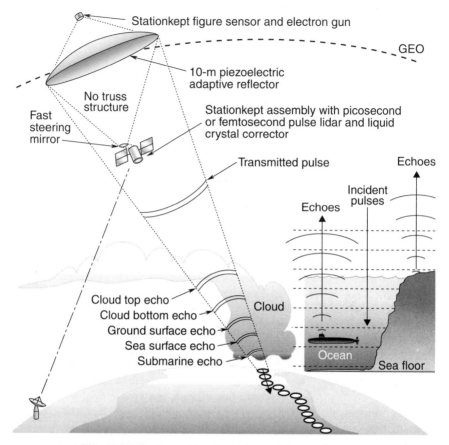

Fig. 15.16. Surface and undersea sensing through clouds.

The required incident power densities are similar to those used for airborne systems and determine the optics size and lidar power. The total weight of this system is about 500 kilograms, based on scaling airborne lasers and the use of adaptive thin film optical reflectors. (If implemented using Buckytubes throughout, its total weight would be reduced in the far term to about 5 kilograms.)

15.16.5 Technologies and Time Frame

The principal technologies required are those of piezoadaptive membranes corrected by electron beams, liquid crystals with large correction capability, MEMS FEEP thrusters for stationkeeping, and space qualifying picosecond pulse lidars. These technologies probably could be demonstrated by 2010 and deployed by 2015, making this a midterm concept.

15.17 Minefield and Mine Detection Spacecraft

15.17.1 Concept Purpose and Utility

Techniques already possible from aircraft but having the coverage and relative immunity of space systems are used to detect from space buried mines and minefields over large areas anywhere on the globe. Implementation would relay Global Positioning System coordinates of detected mines to ground crews for disablement or to air crews or standoff weapons for destruction. The system is shown in Fig. 15.17.

15.17.2 Principles of Operation

The signature of turned Earth in the visible and infrared light wavelengths is compared to that of undisturbed Earth. This technique appears to be useful for change detection of mine positions for up to 1–2 years from mine deployment. To be used from space, a large aperture sensor is needed to attain a resolution sufficient to discriminate and locate adjacent mines.

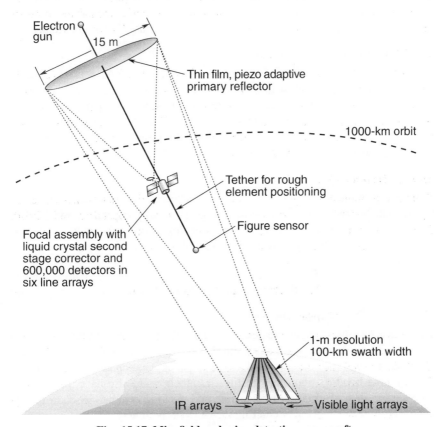

Fig. 15.17. Minefield and mine detection spacecraft.

An adaptive piezoelectric membrane reflector, adjusted by an electron beam in response to a figure sensor, is used as the primary aperture, and a liquid crystal modulator corrects the remaining errors. A focal array having many detectors in a number of lines is used for each of two wavelengths. The array forms a push broom sensor, which is swept across the desired area either by spacecraft motion or by attitude and translation control of a stationkept fast steering mirror, or both. The detected location of the mines is translated into GPS coordinates and transmitted to the ground.

15.17.3 Characteristics of the Illustrated System

A 15-meter-diameter optical system is implemented using an adaptive membrane first stage corrector and a liquid crystal second stage corrector. All elements are stationkept, and there are no structural trusses. A tether may be used to hold the coarse positions of the elements with respect to each other, or they may be self controlled using MEMS FEEP thrusters on the elements. The focal plane contains six line arrays of 100,000 detectors each, half in the visible and half in the infrared portions of the spectrum. The system is placed into an orbit of 1000–kilometer altitude.

15.17.4 Performance and Weight

A 1-meter ground resolution is attained in the infrared at 5-micron wavelength, with a 100-kilometer swath width. One spacecraft suffices for global detection of minefields. The number of spacecraft employed is scenario dependent and may increase depending on the frequency of newly installed mines or the time lag permissible for their detection. The system will detect most mines that are not under canopy and were buried up to 2 years ago.

15.17.5 Technologies and Time Frame

Principal technologies required are those of piezoadaptive membranes corrected by electron beams, liquid crystals with large correction capability, and MEMS FEEP thrusters for stationkeeping. These technologies probably could be demonstrated by 2010 and deployed by 2015, making this a midterm concept.

15.18 Passive Aircraft Surveillance

15.18.1 Concept Purpose and Utility

This concept enables the passive detection of unlimited numbers of large subsonic aircraft, whether conventional or stealth, in the regions around the periphery of CONUS. Its implementation would have principal application as a detection fence for incoming bombers and other hostile aircraft. It could also have a peacetime function for air traffic control in detecting and tracking commercial and military large aircraft traffic leaving or entering the CONUS periphery.

15.18.2 Principles of Operation

The system detects either the direct infrared radiation from the engines or engine plumes of high flying large aircraft, or it detects the "hole" in the normal Earth background caused by the presence of the aircraft. Because of this ability the system detects both conventional aircraft and stealth aircraft. Detection occurs only at high altitudes because of the lower absorption of the atmosphere in the signal path to the spacecraft. A large aperture infrared sensor is used for detecting either the direct radiation or the reduction of background radiation. An implementation of the concept is shown in Fig. 15.18.

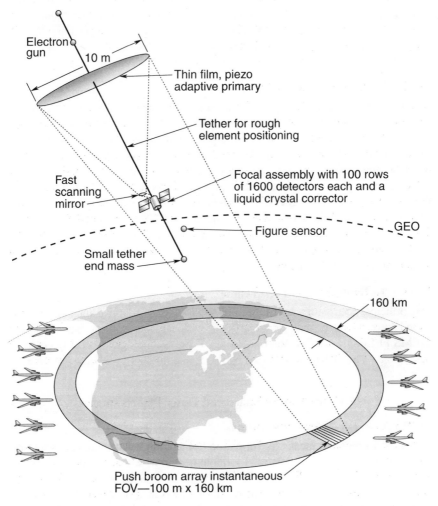

Fig. 15.18. Passive aircraft surveillance.

15.18.3 Characteristics of the Illustrated System

The particular implementation in Fig. 15.18 places the sensor spacecraft in GEO and sweeps a circular detection fence around the periphery of the United States, looking for telltale signatures. The concept employs a 10-meter-diameter optical aperture adaptive piezoelectric membrane in the near infrared. A separated focal assembly containing a liquid crystal corrector and focal plane, a figure sensor, and an electron beam generator are used, which are roughly positioned using a tether and finely positioned using MEMS FEEP thrusters on all the elements.

There is no truss structure. A stationkept fast steering mirror is moved in a circular pattern around the focal assembly, sweeping the field of view in a circular pattern around the periphery of the United States. The sensor implements a push broom array with instantaneous coverage of 100 meters by 160 kilometers, sweeping the CONUS every minute, and designed to detect unlimited numbers of subsonic aircraft. The focal plane consists of an array of 100 rows of 1600 detectors each, passively cooled.

15.18.4 Performance and Weight

A reliable detection sensitivity of 750 watts per steradian is attained, sufficient for detecting large aircraft either by direct radiation or the absence of background signals. Each aircraft could be tracked to a positional accuracy of 100 meters every minute. Low observable aircraft designed to reduce cross sections from the forward or side aspects so as to penetrate ground defenses would find that their detectability is governed rather by their physical cross section as seen from above, thus negating their low observable efforts. At least 400 aircraft can be tracked simultaneously with an update rate of 1 minute, or 4000 with an update rate of 10 minutes. The entire spacecraft would weigh less than 200 kilograms in GEO. (If implemented using Buckytubes throughout, its total weight would be reduced in the far term to about 2 kilograms.)

15.18.5 Technologies and Time Frame

The principal technologies required are those of piezoadaptive membranes corrected by electron beams, liquid crystals with large correction capability, and MEMS FEEP thrusters for stationkeeping. These technologies probably could be demonstrated by 2010 and deployed by 2015, making this a midterm concept.

15.19 Distributed Spacecraft Outer Planets Imager

15.19.1 Concept Purpose and Utility

This concept presents means for formation flying a large distributed astronomical telescope system in the midterm. Its implementation would be able to resolve Kuiper belt objects as well as obtain high resolution images of the moons of Neptune and Uranus. The concept is the result of a dedicated design study at NASA's Jet Propulsion Laboratory commissioned by the author while he was with NASA, and is illustrated in Fig. 15.19.

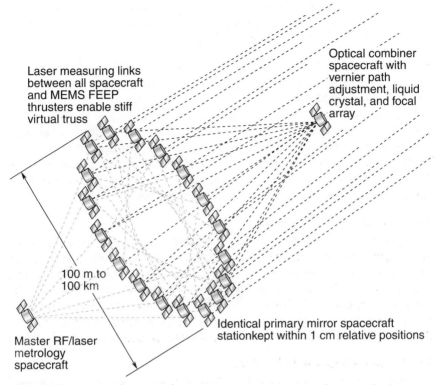

Laser measuring links
between all spacecraft
and MEMS FEEP
thrusters enable stiff
virtual truss

Optical combiner
spacecraft with
vernier path
adjustment, liquid
crystal, and focal
array

100 m to
100 km

Master RF/laser
metrology
spacecraft

Identical primary mirror spacecraft
stationkept within 1 cm relative positions

Fig. 15.19. Distributed spacecraft outer planets imager.

15.19.2 Principles of Operation

A number of identical optical spacecraft are arranged in a circular pattern in space, with each spacecraft acting as a relay, reflecting light from a target to a central collection and combining spacecraft. The light paths from each relay spacecraft to the collection spacecraft are adjusted by precision stationkeeping and by vernier optical path control in the combiner so that they all add in phase. The separation between the relay spacecraft, and thus the resolution of the sparse optical array, is determined by the diameter of the ring in space, while the sensitivity is determined by the sum of the aperture areas.

A metrology spacecraft is stationkept behind the array, establishing laser and RF links to all spacecraft. Each relay spacecraft also establishes such metrology links to some or all other relay spacecraft. These links are used to position the spacecraft relative to each other using propulsion aboard each, and thus establish the equivalent of long stiff virtual trusses between all spacecraft, with the constellation moving as one structure without the weight of a physical truss. The constellation is moved as an ensemble for repointing, tracking, and rotation. This measurement and positioning technique attains a relative accuracy of 1 centimeter among all spacecraft. Variable length optical time delay mechanisms and liquid crystals within the combiner spacecraft in the optical path from each relay spacecraft result in phase coherent addition of all relayed signals.

15.19.3 Characteristics of the Illustrated System

The illustrated system has 16 relay spacecraft, each with a 1-meter-diameter adaptive film mirror and MEMS FEEP propulsion to effect attitude and translation control, but no detectors or processors and are thus very lightweight and simple. The constellation diameter can be varied from 100 meters to at least 100 kilometers by command. The formation flying system attains stationkeeping accuracies of 1 centimeter rms, and the optical trombone and liquid crystal systems in each of the 16 light paths in the combiner spacecraft correct the remainder and attain path accuracies of nanometers for optical coherence and signal addition.

15.19.4 Performance and Weight

The system would allow good resolution imaging in a snapshot mode on moderately bright targets, though a 180-degree rotation of the ensemble about its optical axis would be needed to attain good resolution for very dim targets. The system could obtain 100–1000 pixel images of the moons of Neptune and Uranus and 10–100 pixel images of many Kuiper belt objects. The entire ensemble of spacecraft would weigh less than about 500 kilograms. (If implemented using Buckytubes throughout, its total weight would be reduced in the far term to about 5 kilograms.)

15.19.5 Technologies and Time Frame

The principal technologies required are those of adaptive membrane 1-meter mirrors, liquid crystals with large correction capability, and MEMS FEEP thrusters for stationkeeping. These technologies probably could be demonstrated by 2010 and deployed by 2015, making this a midterm concept.

15.20 Large, Lightweight, Long Dwell Ground Imager

15.20.1 Concept Purpose and Utility

A long dwell or continuous dwell ground imaging capability from GEO with medium to high resolution at extremely low weight and cost would overcome the weaknesses associated with known overflight times of low altitude spacecraft. The value of such a capability would be inestimable. The concept is illustrated in Fig. 15.20.

15.20.2 Principles of Operation

A large diameter optical system attains desired resolution performance from GEO, which location results in the desired long dwell capability. A two-stage correction system is used. The first stage uses an adaptive piezoelectric membrane actuated by an electron beam in response to an optical figure sensor to implement a lightweight primary. The focal assembly contains a second stage of correction using a

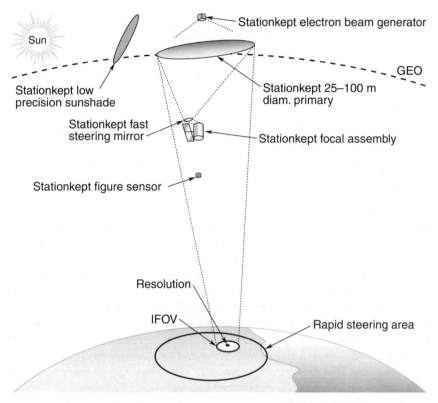

System characteristics for three apertures in GEO

Diameter (m)	Total Weight (kg)	Resolution (m)	IFOV (km)	Rapid Point Field (km)
25	260	1.7	5.0	500
50	600	0.8	2.5	250
100	1800	0.4	1.2	125

Fig. 15.20. Large, lightweight, long dwell ground imager.

liquid crystal spatial light modulator to remove the residual errors on the primary. There is no truss structure. The field of view is swept rapidly by a stationkept small mirror that relays the light from the primary to the focal assembly while maintaining constant path length. The primary is shaded from direct sunlight by a roughly flat stationkept film sunshade. A similar but astronomy application concept was defined by the author in a study for the NASA Institute for Advanced Concepts (see Reference 3.4).

15.20.3 Characteristics of the Illustrated System

The system characteristics were calculated for an aperture of 25 meters in GEO, and scaled to an aperture of 100 meters in diameter. The adaptive membrane primary is deliberately unsupported and without tension, and its attitude and translation control are attained via MEMS FEEP thrusters. The figure sensing, electron beam, steering mirror and focal assembly elements are stationkept with respect to the aperture. Each element of the system was designed so that the system operates as a complete ground imaging system, and its total weight estimated. The stationkept fast steering mirror is used for pointing changes within a ground theater, while the entire ensemble is rotated to point at another theater within the visible hemisphere. The sunshade maneuvers continuously so as to interpose itself between the sun and the primary aperture.

15.20.4 Performance and Weight

The two-stage optical system design is capable of attaining near diffraction limited performance with a minimum on orbit operation time of five years (see Reference 4.1). The designed system can attain a 1.7-meter ground resolution from GEO with a 25-meter primary aperture and an instantaneous field of view of 5 kilometers. This field can be moved 500 kilometers in a fraction of a minute by changing the location and attitude of only the fast steering mirror. The entire ensemble of elements can be repointed anywhere within 60 degrees on the Earth within 15 minutes and attain continuous dwell. The scaled up system can attain resolutions of 0.34 meters from GEO with a 100-meter primary and an instantaneous field of view of 1.2 kilometers. This field can be moved 125 kilometers with the steering mirror alone anywhere within 60 degrees on the Earth by rotating the ensemble.

The 25-meter imaging system would weigh 260 kilograms total in GEO, while the 100-meter system would weigh 1800 kilograms in GEO. (If implemented using Buckytubes throughout, the 25-meter system in GEO would weigh 3 kilograms, while the 100-meter system in GEO would weigh 18 kilograms.)

15.20.5 Technologies and Time Frame

The principal technologies required are those of piezoadaptive membranes corrected by electron beams, liquid crystals capable of large corrections, and MEMS FEEP thrusters for stationkeeping. These technologies probably could be demonstrated by 2010 and deployed by 2015, making this a midterm concept.

15.21 Large, Lightweight, Long Dwell, Tethered Ground Imager

15.21.1 Concept Purpose and Utility

The description of this concept is similar to that of the preceding related concept, with the exception that many of the principal elements are supported by a long tether to minimize the propulsive requirements of stabilizing and operating it. Its purpose and utility are the same as for the preceding concept. See Fig. 15.21.

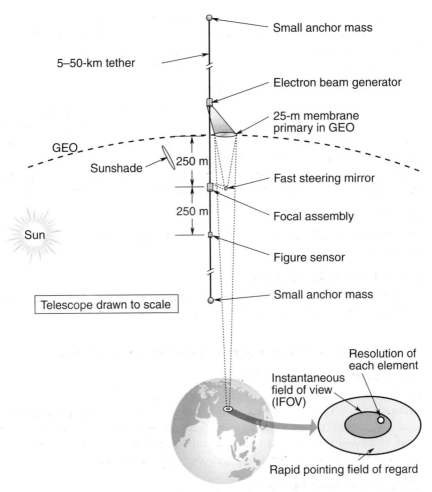

System characteristics for three apertures in GEO

Diameter (m)	Total Weight (kg)	Resolution (m)	IFOV (km)	Rapid Point Field (km)
25	260	1.7	5.0	500
50	600	0.8	2.5	250
100	1800	0.4	1.2	125

Fig. 15.21. Large, lightweight, long dwell tethered ground imager.

15.21.2 Principles of Operation

Most of the elements of the ground imager lie nearly in line and nearly along the local vertical. Thus a tether can be effectively used to hold these elements against the Earth's gravity gradient as opposed to continuous thrusting propulsion on the separated elements, which saves substantial amounts of propellant. The tether will be lighter than the propulsion systems and propellants, though propulsion will still be required on all elements but only for vernier control.

15.21.3 Characteristics of Illustrated System

The tether used is up to 50 kilometers long. It is anchored at the ends by 10-kilogram masses and in GEO by a smaller mass, all of which have active MEMS FEEP propulsion. Anchors provide the stabilizing tension and also remove the effects of tether dynamics. The sunshade element is moved propulsively half of its diameter twice a day to avoid colliding with the tether, but the extra propellant for these maneuvers is small compared to that required to resist sunlight pressure.

15.21.4 Performance and Weight

The performance and weight of this system are the same as those of the previous, untethered system.

15.21.5 Technologies and Time Frame

The technologies are the same as those of the previous concept, with the addition of a lightweight, long-life tether. The time frame is also the same.

15.22 Large, Lightweight, Astronomical Space Telescope

15.22.1 Concept Purpose and Utility

The concept shown in Fig. 15.22 is also similar to that of 15.20, except that it is optimized for deep space operation and detection and imaging of astronomy objects. It presents means for implementing an astronomical telescope with much larger collection aperture and finer resolution than NASA's James Webb Space Telescope (JWST), yet with much lower weight and likely cost.

15.22.2 Principles of Operation

The principles are similar to those of concept 15.20 with some significant differences: the sunshade stationkeeping is simpler and its rotation much slower, the system is placed into solar orbit perhaps at a Lagrangian point or in deeper space requiring less propellant to maintain, passive cooling of the system is readily attained for infrared operation, and dedicated shields can be interposed out of the field of view between the spacecraft and celestial objects being viewed to screen out very bright stars.

15.22.3 Characteristics of the Illustrated System

The system is an "outward" pointed version of those described in the previous two concepts. Its elements and their functioning are very similar.

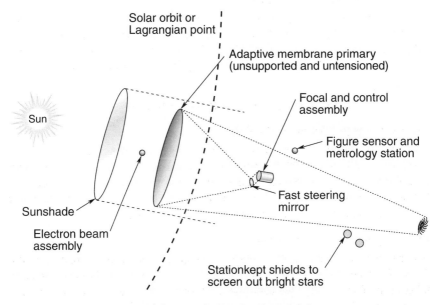

System characteristics for three apertures in solar orbit

Diameter (m)	Total Weight (kg)	Resolution (microradians)	IFOV (milliradians)
25	260	0.04	0.12
50	600	0.02	0.06
100	1800	0.01	0.03

Fig. 15.22. Very large, lightweight, astronomical space telescope.

15.22.4 Performance and Weight

The designed system can attain a resolution of 40 nanoradians with a 25-meter primary aperture with an instantaneous field of view of 0.12 milliradians, and a collecting area 10 times greater than JWST. The scaled system can attain resolutions of 10 nanoradians with a 100-meter primary with an instantaneous field of view of 0.03 milliradians and a collecting area 156 times greater than JWST, thus being able to detect faint objects 12.5 times further away. The 25-meter system would weigh 260 kilograms in orbit; whereas the 100-meter system would weigh 1800 kilograms. (If implemented using Buckytubes throughout, the 25–meter system would weigh 3 kilograms; the 100-meter system would weigh 18 kilograms.)

15.22.5 Technologies and Time Frame

The principal technologies required are those of piezoadaptive membranes corrected by electron beams, liquid crystals with large correction capability, and MEMS FEEP thrusters for stationkeeping. These technologies probably could be demonstrated by 2010 and deployed by 2015, making this a midterm concept.

15.23 Zooming Long Dwell Ground Imager

15.23.1 Concept Purpose and Utility

This concept presents a long dwell or continuous dwell ground imaging capability from GEO with medium-high resolution at extremely low weight and cost, with an additional ability to zoom in on desirable targets in real time. Its implementation would allow one imaging system to function both as a finder as well as a high resolution imager system with an uninterrupted capability.

15.23.2 Principles of Operation

A large diameter, sparse aperture optical system is implemented to make possible the resolution performance from GEO, which ensures the desired long dwell capability. A number of individual apertures are formed into a ring. Each consists of an adaptive piezoelectric membrane, actuated by electron beam in response to an optical figure sensor. Each aperture is deliberately unsupported and without tension; its attitude and translation control are attained via MEMS FEEP thrusters.

The focal assembly contains a second stage of correction for each primary aperture using liquid crystal spatial light modulators and the detector array receiving the in-phase combined signal. The apertures are positioned to form a sparse ring whose diameter can be commanded to adjust for zooming. The figure sensing, electron beam, and focal assembly elements are coarsely positioned using a very lightweight tether, and finely positioned with respect to the apertures by MEMS FEEP thrusters on all the elements. All the separation distances are adjusted when zooming. The field of view is swept rapidly by a stationkept small mirror that relays the light from all the primaries to the focal assembly. There is no truss structure.

15.23.3 Characteristics of the Illustrated System

There are 24 apertures, each with a 4-meter diameter. The system can zoom from 60 to at least a 200-meter overall diameter. The ensemble is repointed at another theater by rotating it to image another area within the visible hemisphere. The tether is actively positioned along the optical axis of the system. (See Fig. 15.23.)

15.23.4 Performance and Weight

The two-stage multiple path optical system will attain near diffraction limited performance. The designed system can attain a 70-centimeter resolution in a 3-kilometer instantaneous field of view when the apertures are in a 60-meter diameter,

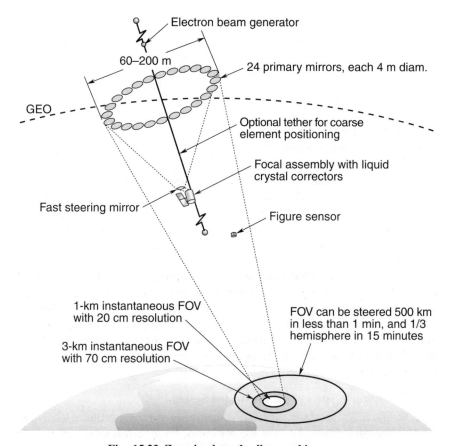

Fig. 15.23. Zooming long dwell ground imager.

and can be continuously zoomed to a 20-centimeter resolution in a 1-kilometer field of view when in a 200-meter diameter, all from GEO. These fields of view can be moved 500 kilometers by changing the location and attitude of only the fast steering mirror within a fraction of a minute, and can be repointed anywhere within 60 degrees on Earth within 15 minutes. The total system would weigh less than about 200 kilograms in GEO. (If implemented using Buckytubes throughout, its total weight would be reduced in the far term to about 2 kilograms.)

15.23.5 Technologies and Time Frame

The principal technologies required are those of piezoadaptive membranes corrected by electron beams, liquid crystals having large correction capability, and MEMS FEEP thrusters for stationkeeping. These technologies probably could be demonstrated by 2010 and deployed by 2015, making this a midterm concept.

15.24 Ultraresolution Sparse Aperture GEO Imager

15.24.1 Concept Purpose and Utility

With this concept, a low weight space system yields exceedingly high surface resolution images from GEO resulting in simultaneous large coverage, long dwell, and fine resolution. Implementation of this concept would overcome the weaknesses associated with known overflight times of low altitude spacecraft. The value of such a capability would be inestimable.

15.24.2 Principles of Operation

This concept illustrates another form of a sparse GEO imager, similar in principle to that of the previous concept, however in this case five sparse aperture rings in space are used simultaneously. Each ring consists of 24 apertures, each having a 4-meter diameter, and each adaptively corrected by electron beam in response to an optical figure sensor. All elements are stationkept, and a central tether for coarse element positioning may or may not be found advantageous. Fast steering of the instantaneous field of view is attained by a small mirror, stationkept in front of the focal assembly.

The rings can be used independently to simultaneously image five areas, which may or may not be contiguous, by independent control of each aperture position, orientation, and shape. Additionally, even though each optical system is independent of the other, the correcting and combining assembly allows for their simultaneous coherent use to image a smaller area but with a finer resolution given by the overall diameter of the total ensemble. The ensemble can be reoriented for repointing to other areas.

15.24.3 Characteristics of the Illustrated System

Five groups of 24 apertures each are contained in the system illustrated in Fig. 15.24, although the optical design was performed for up to seven such groups. The overall diameter of the assembly is 300 meters, but could be expanded to 500 meters or larger at the cost of image degradation due to sparseness. There are five groups of 24 liquid crystal assemblies in the focal assembly. The focal assembly would be located at 10 times the overall diameter for an f/10 design so that the assembly is stationkept at 3 kilometers below the apertures. All elements are positioned using MEMS FEEP thrusters, and have a 5-year propellant load.

15.24.4 Performance and Weight

The finest resolution of the 300-meter-diameter imager is 14 centimeters in a 1-kilometer field of view. The resolution of each of the separate five rings is 70 centimeters in a 5-kilometer field of view. The system coverage area can be repositioned within a minute within 100 kilometers, and within 30 minutes to any point within a large fraction of the visible hemisphere. The system total weight on orbit is about 1500 kilograms. (If implemented using Buckytubes throughout, its total weight would be reduced in the far term to about 20 kilograms.)

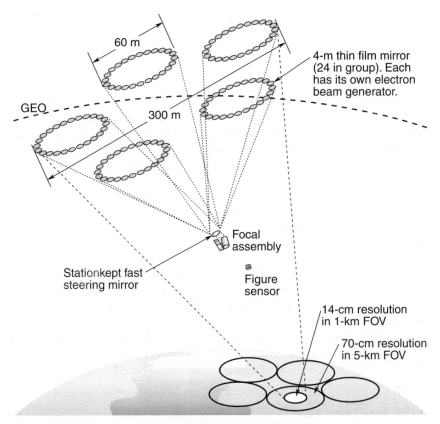

Fig. 15.24. Ultraresolution sparse aperture GEO imager with long dwell capability.

15.24.5 Technologies and Time Frame

The principal technologies required are those of piezoadaptive membranes corrected by electron beams, liquid crystals with large correction capability, and MEMS FEEP thrusters for stationkeeping. These technologies probably could be demonstrated by 2010 and deployed by 2015, making this a midterm concept.

15.25 Ultraresolution Ring Aperture Tethered Ground Imager

15.25.1 Concept Purpose and Utility

A low weight space system achieves exceedingly high surface resolution images from GEO, with large coverage and long dwell. Its implementation would overcome the weaknesses associated with known overflight times of low altitude spacecraft. The value of such a capability would be inestimable.

15.25.2 Principles of Operation

This concept is similar in principle to the two preceding ones. A large diameter but sparse aperture optical system is implemented to make possible the resolution performance from GEO, which ensures the desired long dwell capability. An annular aperture composed of separate shaped segments is formation flown in the ring configuration with edges close but not touching. Each segment is an adaptive piezoelectric membrane, actuated by an electron beam in response to an optical figure sensor, with attitude and translation control attained via MEMS FEEP thrusters.

The focal assembly contains a second stage of correction using a liquid crystal spatial light modulator and the detector array. The figure sensing and electron beam elements are coarsely positioned using a very lightweight tether and finely positioned with respect to the aperture by MEMS FEEP thrusters, which are also present on all elements. The field of view is swept rapidly by a stationkept small mirror that relays the light from the primary to the focal assembly. There is no truss structure. The ring primary is shaded from direct sunlight by a stationkept variable geometry but low accuracy ring sunshade, with segments corresponding one for one to those of the primary.

15.25.3 Characteristics of the Illustrated System

The system characteristics were calculated for an aperture of 250 meters in GEO and for fill factors of 5, 10, and 20 percent of the overall area. The near continuous annular area results in good image quality even though sparse. A tether supports a number of the elements loosely along the local vertical, eliminating the need for constant thrust. The entire ensemble is rotated to point the field of view at another theater to image another area within the visible hemisphere. Every element of the system was designed, allowing the weight of the total system to be estimated.

15.25.4 Performance and Weight

The system, illustrated in Fig. 15.25, achieves near diffraction limited performance. It can attain 17-centimeter resolution from GEO. Its weight depends on the width of the primary ring, which determines the sparseness or fill factor. A 20-percent filled aperture will weigh 2380 kilograms, a 10-percent filled aperture 1620 kilograms, and a 5-percent filled aperture 1140 kilograms. The instantaneous field of view is 500 meters, which can be rapidly moved within a 100-kilometer-diameter area without repointing the elements by the fast steering mirror. The ensemble can be repointed to another area on the globe by rotating the ensemble in under 30 minutes. (If implemented using Buckytubes throughout the total weight of the 20-percent filled system would be 24 kilograms; the 10-percent system, 16 kilograms; and the 5-percent system, 11 kilograms in the far term.)

15.25.5 Technologies and Time Frame

The principal technologies required are those of piezoadaptive membranes corrected by electron beams, liquid crystals with large correction capability, long-life tethers,

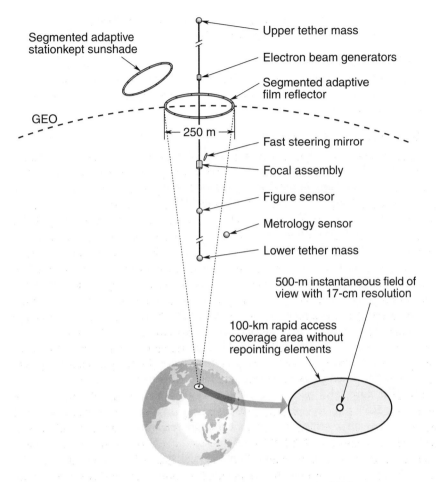

Fig. 15.25. **Ultra resolution ring aperture tethered GEO imager with long dwell capability.**

and MEMS FEEP thrusters for stationkeeping. These technologies probably could be demonstrated by 2010 and deployed by 2015, making this a midterm concept.

15.26 Interferometer to Image Earth-Size Planets Orbiting Stars

15.26.1 Concept Purpose and Utility

A large, low weight interferometer images Earth-size planets orbiting nearby stars. Its achievement could address questions of life elsewhere in the universe at an affordable budget.

15.26.2 Principles of Operation

A large diameter, but sparse aperture, optical interferometer system is implemented to make possible the required resolution and performance from solar orbit. The apertures are large in order to collect sufficient photons for the detection of the weak signals expected from such a planet. The interferometer could be either of the Michaelson or the Fizeau type. The multiple large apertures are arrayed in a large ring. Light from each is coherently combined in pairs in the combiner spacecraft for the Michaelson configuration and together for the Fizeau configuration.

A laser information link is established between each pair of apertures for this combination, using a coherent laser reference. Laser metrology links are established between all combiners, one per aperture, which serve to control the attitude and location of each element using MEMS FEEP thrusters as well as carry the combining information. This control creates a stiff virtual truss structure. The system can operate in the infrared or visible regions, as the primaries will be passively cooled by their sunshades being in deep space. An RF pair metrology system determines coarse relative positioning for cueing the laser system.

The laser links also operate as kilometric gyros to measure the rates of the ensemble. Each primary aperture is an adaptive piezoelectric membrane, actuated by electron beam in response to an optical figure sensor, with attitude and translation control attained via MEMS FEEP thrusters. Each focal assembly contains a second stage of correction using a liquid crystal spatial light modulator and the focal detectors. Each primary has an associated stationkept sunshade and thermal conditioner. There is no truss structure.

15.26.3 Characteristics of the Illustrated System

The illustrated system in Fig. 15.26 consists of 25 primary apertures, each 25 meters in diameter, arrayed in a ring with a 250-kilometer diameter. The entire ensemble is rotated normal to its axis of symmetry to point the field of view at another area in the celestial sphere.

15.26.4 Performance and Weight

The interferometer system can resolve 100 to 300 pixels on an Earth-size planet around a nearby star. The total weight of the system in space is about 3500 kilograms. (If implemented using Buckytubes throughout, its total weight would be reduced in the far term to about 35 kilograms.)

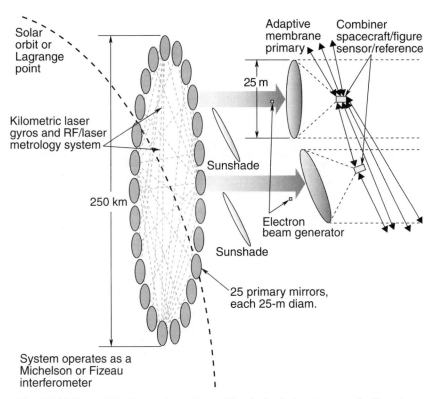

Fig. 15.26. Large interferometer to image Earth-sized planets around other stars.

15.26.5 Technologies and Time Frame

The principal technologies required are those of piezoadaptive membranes corrected by electron beams, liquid crystals with large correction capability, MEMS FEEP thrusters for stationkeeping, kilometric laser gyros, and laser coherent combiners. These technologies probably could be demonstrated by 2015 and deployed by 2020, making this a far term concept.

16 Transportation and Infrastructure Concepts

These concepts address access to space, transportation within space, and the space facilities, structures, and infrastructures needed for support of ambitious undertakings in space. Concepts that enable operational functions are the focus of this chapter, but some fairly generic applications that could not be characterized as radio frequency or optical are also included.

The 18 concepts in this category are organized as in the preceding chapters in Part IV and are listed in Table 16.1 according to the near term, midterm, and far term time frame in which they are likely to be attainable. Within that classification, they are presented in approximate order of increasing complexity, longer time frame, or both. All these concepts are unclassified, and each is illustrated and discussed in the order listed.

Table 16.1. Transportation and Infrastructure Concepts

Time Frame	Concept
Near term	Avoiding human degradation from zero gravity during long space flights
	Increasing performance and decreasing cost of reusable launch vehicle payload deployment
	Spacecraft altitude and inclination change without using propellants
	Battery replacement via orbital accumulator
	Satellite defensive countermeasures
Midterm	Earth-stationary "polesitter" spacecraft
	Aero-gravity assist for interplanetary spacecraft
	Global range gun/orbital accelerator
	Electromagnetic launch assist
	Hard and submerged target kill from space
	Combination propulsion system single stage to orbit (SSTO) launch vehicle
	Fast LEO to GEO transportation without using propellants
	Earth-to-moon transportation system requiring no propellants
Far term	Mars Phobos-Deimos ladder
	Nuclear waste disposal
	Radio frequency powered electromagnetic LEO launch system
	Geostationary low altitude spacecraft
	Macroengineering Earth-to-space transportation concepts

16.1   Avoiding Human Degradation From Zero Gravity

16.1.1 Concept Purpose and Utility

An Earth-level gravity environment for long duration human spaceflight, such as missions to the planets, would enable space voyages to be arbitrarily long without subjecting the crew or passengers to the deleterious effects of the absence of gravity forces. Because these effects would be avoided, this capability would reduce the ongoing costly expenditures in life sciences research into the long-term effects of zero g on humans. Furthermore the presence of substantial g forces allows more efficient operations, much akin to the usual Earth-bound activities to which humans are accustomed. This concept addresses only the g force effects and not the radiation effects of long duration space flight.

16.1.2 Principles of Operation

The space vehicle is separated into two parts, connected by a short tether and slowly spun to rotate about each other. The tether allows spin radii large enough so as to result in normal Earth g levels without any coriolis force disorientation. The tether length can easily be made a kilometer or more, since 20-kilometer tethers have already been successfully deployed in space. With such separation lengths, the rotation rate to produce 1 g in the crew facility attached to the tether end is one revolution per minute or less, a rate that experts agree will produce no space sickness. Furthermore the gradient between the head and feet is so small that the human is comfortable in all orientations and motions. The tether can be a multiline design such as a "Hoytether" caduceus design, which is constructed of multiple interconnected strands and will last decades in the space debris environment. Tether length and rotation rate can be easily varied in flight to change the g level felt at the facility.

16.1.3 Characteristics of the Illustrated System

An initial tether length of 1000 meters could be used with the modules at the spacecraft ends spun to one revolution per minute using small rockets, resulting in 1 g at the modules. Longer tethers can be used, such as a 4000-meter tether, which would yield 1 g at 0.5 revolutions per minute. Lower g levels can be created by using shorter tethers at the same rotation velocities, or longer tethers at slower rotation velocities. Conversely, higher g levels could also be generated if found to be desirable. The length of a tether can be varied in flight by reeling, thus changing the rotation speed with the angular momentum remaining constant, which results in changing g levels without energy input other than the reeling power, which could be mostly recovered for repetitive operations. (See Fig. 16.1.)

16.1.4 Performance and Weight

For Mars missions, crews can leave Earth with a spin rate resulting in 1 g, slowly decrease the spin rate to 3/8 g by paying out tether, and arrive at Mars fully adjusted to its 3/8 g. On the return trip the g level could start at 3/8 g and gradually increase to 1 g near Earth. The crews are fit to work at all times, never ill, and their

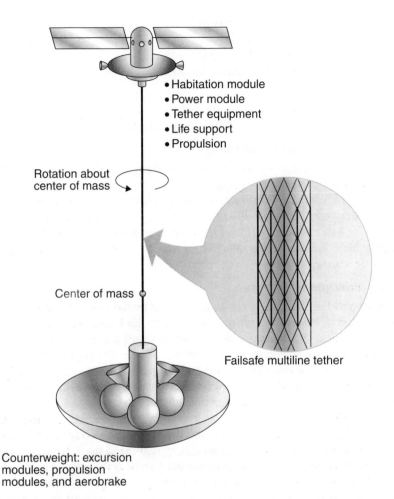

- Habitation module
- Power module
- Tether equipment
- Life support
- Propulsion

Rotation about center of mass

Failsafe multiline tether

Center of mass

Counterweight: excursion modules, propulsion modules, and aerobrake

Tether Length to Center of Mass (m)	RPM	G's at Module
1000	1	1
375	1	3/8
4000	0.5	1
1500	0.5	3/8

Fig. 16.1. A space vehicle in two parts connected by a tether rotates to create normal Earth g levels to avoid human degradation during long space flights.

bodies do not deteriorate. The tether is lightweight and inexpensive. (If implemented using Buckytubes throughout, its weight in the far term would be minuscule.)

16.1.5 Technologies and Time Frame

The only technology required is that of fail-safe multiline tethers or ribbon-type tethers, either of which can have multidecade lifetimes. The tether portion could be deployed well before 2010 and is thus a near term concept.

16.2 Increasing Performance and Decreasing Cost per Kilogram of any Launch Vehicle

16.2.1 Concept Purpose and Utility

A simple tether can be used to increase the payload capability of any launch vehicle while simultaneously decreasing the cost of placing that payload into its orbit. It works on any launch vehicle and requires few modifications. Its implementation would effectively add capability at a lower marginal cost per kilogram than the cost of the basic vehicle itself. Though most effective on fully reusable and single stage vehicles this concept is effective on any vehicle regardless of its number of stages or whether it is reusable or expendable.

16.2.2 Principles of Operation

As shown in Fig. 16.2, the payload is deployed by the usual spring release after reaching orbit; however, it remains tethered to the launch vehicle. The launch vehicle places its last stage (or itself if single stage) and the payload and tether into a lower and more eccentric orbit than that desired. The payload is then slowly extended upward by unreeling the tether. This process transfers energy and angular momentum from the launch vehicle mass to the payload mass, with the sum of both remaining constant. At the right time, the tether is cut.

The payload's excess energy causes it to go into a higher apogee orbit, and the identical decrement in energy causes the launch vehicle to enter a lower perigee orbit. Thus energy and momentum have been scavenged from the launch vehicle (which can reenter directly if desired) and transferred to the payload to increase its orbital energy, at nearly 100-percent efficiency. With proper choice of initial orbits, the payload enters its desired higher energy orbit without the launch vehicle expending any further propellants and without an upper or kick stage. This is the closest thing to a free lunch. Furthermore, the tether weighs much less that an upper stage to accomplish the same effect, and costs far less.

16.2.3 Characteristics of the Illustrated System

System designs have been carried out for payloads launched by a Delta-class staged expendable launch vehicle as well as for an SSTO launch vehicle design with payload capabilities similar to those of the Space Shuttle. In either case the tether system deploys the payload from an initially elliptic orbit into the desired circular orbit by simple tether release. The fully reusable vehicle has the greater benefit owing to the larger mass of the final orbital stage, which can transfer more energy and angular momentum to a given mass payload. Payload altitudes can be increased, inclinations changed, or both using this technique.

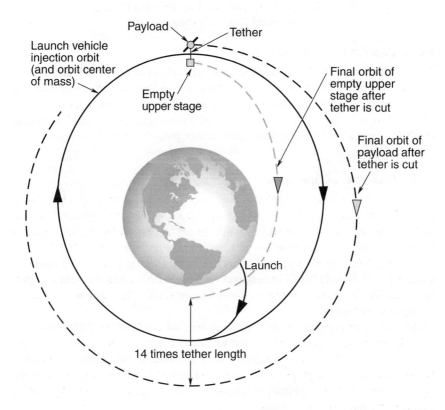

Fig. 16.2. Using a tether to deploy payload in space from a launch vehicle, increasing its performance and decreasing the cost per kilogram.

16.2.4 Performance and Weight

Calculations for an existing launch vehicle, the Delta III, which can place about 6000 kilograms into a 1000-kilometer polar orbit, show that the use of a tether results in a 280-kilogram gain in payload weight into the same orbit. Subtracting the deployer and tether weight leaves a net 180-kilogram gain. Since the cost of the tether equipment is expected to be only $200,000, the cost of the extra capability gained is only $1300 per kilogram, which is an order of magnitude lower than the cost of the basic Delta III launch vehicle.

16.2.5 Technologies and Time Frame

There are no new technologies to demonstrate, as all aspects have already been space tested. The system concept has simply not been embraced yet. Thus the system could be deployed between 2005 and 2010, making it a near term concept.

16.3 Spacecraft Altitude and Inclination Change Without Using Propellants

16.3.1 Concept Purpose and Utility

A spacecraft can be made to maneuver in space without requiring any propellants. Implementation of this concept would allow raising the orbit to compensate for drag, reducing the orbit altitude to cause a spacecraft to deorbit at the end of its life, or change the inclination or any other orbit parameters in any combination, at much lower cost and weight than using rocket propulsion. The system is illustrated in Fig. 16.3.

16.3.2 Principles of Operation

A reversible electromagnetic tether is implemented on a spacecraft that has a solar array. Current from the solar array is fed into the tether when the spacecraft is sunlight illuminated, causing electrodynamic forces generated by the wire cutting Earth's magnetic field at high speed to accelerate the spacecraft—the tether is used in the motor mode and thus raises the orbit. Current is extracted from the tether when it is in darkness in conjunction with Earth's magnetic field and thus operating in the generator mode, lowering the orbit. The current is returned through the ionosphere.

The system will operate at altitudes up to approximately 1000–2000 kilometers because ionospheric plasma is needed to conduct return currents. By using the vector direction of the Earth's magnetic field to modulate the acceleration and/or braking functions at appropriate intervals and orbital locations, any or all of the orbital elements of a low altitude spacecraft can be changed without firing a single engine or burning any propellants.

16.3.3 Characteristics of the Illustrated System

The system has a 10-kilometer long conducting tether. It can be mostly bare wire to increase the collecting area for ionospheric charges, with a short insulated section at the spacecraft end to insulate it from the spacecraft itself and bleed currents. A plasma contactor or field emission contactor enables closing the current loop. The cross section of the wire is matched to the current and thrust desired, and for many applications will be lightweight and inexpensive.

16.3.4 Performance and Weight

Many orbital altitude changes can be effected in days, and large inclination changes can be done in weeks, all with no propellants. Specific numbers can be readily developed for specific goals. This system has been demonstrated in principle in space in NASA's Plasma Motor Generator program at low power, and will be demonstrated at higher power in the next year by NASA's Propulsive Small Expendable Deployer System program.

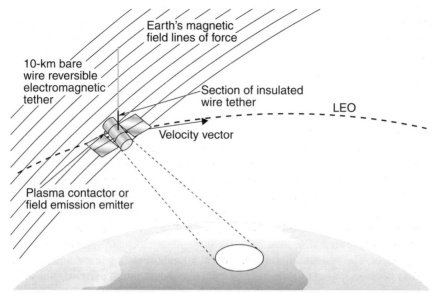

Motor mode
- Portion of solar array power drives current up the wire.
- Current returns through the plasmasphere and interacts with Earth's magnetic field, producing an accelerating force.
- This force raises altitude, changes inclination, or cancels drag.

Generator mode
- Uses self-induced voltage to pass a current down wire through payload.
- This creates a decelerating force that produces power at the expense of altitude.

Fig. 16.3. System concept for changing spacecraft altitude and inclination without using propellants.

16.3.5 Technologies and Time Frame

The only technologies required are those of high power long-life tethers and plasma contactors or other means of coupling current into the plasmasphere. These technologies could be demonstrated well before 2010 and deployed well before 2015, making this a near term concept.

16.4 Orbital Accumulator Battery Replacement

16.4.1 Concept Purpose and Utility

This concept presents a 100-percent depth of discharge, almost 100-percent efficient battery equivalent system with indefinitely large numbers of cycles, no moving parts, and long life. It can also provide high electrical power for short times using a small solar array by acting as a direct-current power transformer. Its implementation can thus replace batteries on most spacecraft and act as a high power intermittent power source at little expense compared with a conventional solar array power system.

16.4.2 Principles of Operation

A reversible electromagnetic tether is implemented on a spacecraft that has a solar array but no battery. When the spacecraft is illuminated by sunlight, current from the solar array powers the payload as well as feeds the tether, causing electrodynamic force acceleration as the current cuts the Earth's magnetic field at high velocity. The tether is thus used in its motor mode to raise the orbit. Current is extracted from the tether using electrodynamic force deceleration when the spacecraft is in darkness and in the generator mode, while lowering its orbit. The excess array power that would normally be used to charge a battery is thus stored as a small increase in the orbit altitude, or as potential energy. This power is recovered at night as the spacecraft is lowered to the original altitude in the process of generating power. The system will only operate at altitudes below a few thousand kilometers because it needs ionospheric plasma to conduct return currents. Since the concept accumulates energy in orbital altitude rather than in battery chemistry it is thus an orbital accumulator.

16.4.3 Characteristics of the Illustrated System

The process is very much like a battery, except that it is highly efficient, capable of 100-percent depth of discharge with no damage, and capable of an indefinitely large number of cycles as there is no obvious wear-out mechanism. Additionally the tether can operate as a direct-current power transformer since the energy stored in raising the orbit can be extracted at a much greater rate than that at which it was imparted. The concept shown in Fig. 16.4 is only notionally designed.

16.4.4 Performance and Weight

This application is much lighter and cheaper than any battery for the same capacity, is not damaged by total discharge or extremely many cycles, has very long life, and can be advantageous to spacecraft that can tolerate some altitude variations. As an example, using a 1-kilowatt solar array and a 50-percent ratio of sunlight to total orbital time in a 90-minute orbit, the tether system can deliver 1 kilowatt of power for 45 minutes, 10 kilowatts for 4.5 minutes, or 100 kilowatts for 27 seconds.

16.4.5 Technologies and Time Frame

The only technologies required are those of high power tethers and plasma contactors or other means of coupling current into the plasmasphere. These technologies could probably be demonstrated well before 2010 and deployed well before 2015, making this a near term concept.

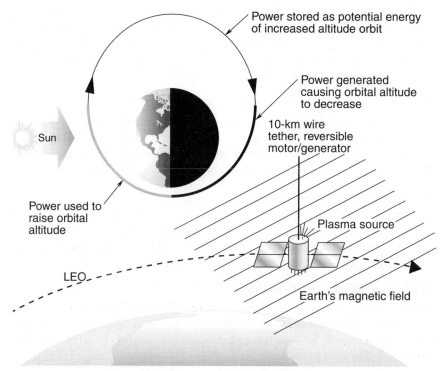

Fig. 16.4. Tethered orbital accumulator used as battery replacement.

16.5 Satellite Defensive Countermeasures

16.5.1 Concept Purpose and Utility

This concept presents one of several possible and novel implementations of defensive countermeasures for spacecraft. Its implementation would prevent successful functioning of incoming interceptors and could be especially useful as countermeasures against staring and home-on-jam interceptors.

16.5.2 Principles of Operation

A spacecraft deploys a set of countermeasures, electronic warfare equipment, and/or decoys on the ends of a set of rotating tethers that have lengths of hundreds of kilometers. The entire ensemble of equipment and tethers is rotated about a pivot point on the spacecraft. The rotational centrifugal forces hold the countermeasure masses at the ends of the tethers in tension by purely passive means.

An incoming interceptor could be foiled by a number of means. The space tracking network could have difficulty determining the spacecraft orbit because countermeasure packages are not in Keplerian orbits. Even if an orbit is determinable, the interceptor trajectory may have to be aimed at the average locations of the object or at the center object, assuming it to be the spacecraft. By proper implementation of electronic warfare techniques, the interceptor aim point could be drawn to a tethered package instead, or the spacecraft could be located on one of the tethers. The tethered packages could be blinked at particularly effective rates to confuse the guidance system. The tethers and the countermeasure equipment could be deployed only in times of impending hostilities.

16.5.3 Characteristics of the Illustrated System

The system illustrated in Fig. 16.5 has four tethers, each of which could be up to 100–200 kilometers long without unduly increasing the system weight. The tethers are attached to the spacecraft as spokes on a wheel so that rotation is possible in a plane. Other possible mounting arrangements could allow cross-track rotation.

16.5.4 Performance and Weight

This system should work well against home-on-jam interceptors and against staring optical or infrared seekers. The potential performance of specific designs against specific designated threats would probably be classified for national security reasons, so this concept is shown only as a notional design.

16.5.5 Technologies and Time Frame

All the technologies needed for this concept have been or will be demonstrated in space in the next year, though not at full scale in the required configuration. Such a demonstration could be done before 2010 and the system deployed before 2015, making this a near term concept.

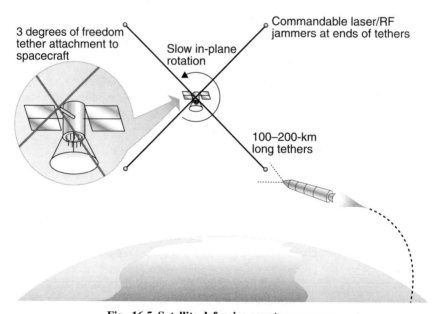

Fig. 16.5. Satellite defensive countermeasures.

16.6 Earth-stationary "Polesitter" Spacecraft

16.6.1 Concept Purpose and Utility

This concept presents the only known practical means to implement a spacecraft that can be geostationary over the Earth's pole or hover over a high Earth latitude without expending energy. Its implementation would be useful for polar communications, continuous sensing of high latitude surface features or aircraft from a single spacecraft, continuous observations of auroral or other polar phenomena, or other applications.

16.6.2 Principles of Operation

A solar sail-equipped spacecraft is placed into a solar orbit, offset from Earth's orbit but matching it in other parameters. The solar pressure push on the spacecraft is counterbalanced by Earth's gravitational pull and the spacecraft hovers over Earth. This balancing of forces requires a location at which the pull of Earth's gravity is comparable in magnitude to the weak push of the solar pressure. This occurs only at distances of many tens of Earth radii from the Earth even for large area solar sails. When over the pole, the spacecraft hovers as does a GEO spacecraft over the equator. When over a high latitude, the spacecraft will appear to move along a constant latitude minor circle as the Earth spins under it. This concept is patented by Robert Forward. Applications such as communications or sensing will require large antenna or sensor diameters to compensate for the long distances from Earth compared with the shorter distance when operating at GEO.

16.6.3 Characteristics of the Illustrated System

The spacecraft illustrated in Fig. 16.6 will be lightweight, consisting principally of thin film reflecting membranes for the sail and tethers by which it pulls the payload. The application apertures could be made very lightweight though large by using the adaptive film techniques described earlier, with a 100-meter-diameter aperture as an example. The principal requirement for the solar sail is a large projected area whose surface need not be an accurate figure, little mass, and a reliable deployment mechanism. Stability analyses were made of a 20-kilometer length heliogyro "polesitter" concept, and indicate that force balance can be attained for such parameters. This implementation is similar to the solar sail proposed for the Halley's comet mission in 1981, but at a different location in space.

16.6.4 Performance and Weight

The spacecraft can hover exactly over the north geographic pole (or south pole) at an altitude of 30 Earth radii in winter and at about 250 Earth radii in summer. Hovering over a constant latitude of 60–70 degrees (north or south) requires that the spacecraft hover at an altitude of 20 Earth radii in winter and 50 Earth radii in summer.

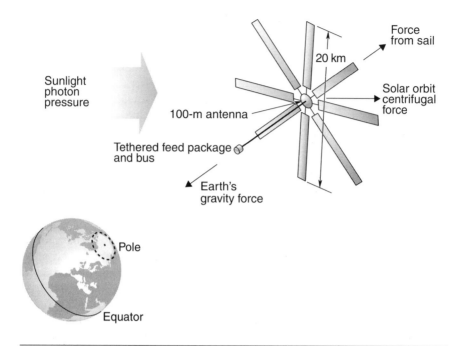

Hover Latitude (deg)	Altitude in Winter	Altitude in Summer
90 (geographic pole)	30 Earth radii	250 Earth radii
60–70 (north or south)	20 Earth radii	50 Earth radii

Fig. 16.6. Earth-stationary "Polesitter" spacecraft.

16.6.5 Technologies and Time Frame

The principal technologies are those of lightweight multiline tethers that survive for decades, solar sail materials that survive in the space environment for long times and can be manufactured in thicknesses of the order of 1 micron or less, and solar sail deployment techniques that assure deployment without tearing or clumping. These technologies could probably all be developed before 2010 and the system deployed before 2015, making this a midterm concept.

16.7 Aero-Gravity Assist for Interplanetary Spacecraft

16.7.1 Concept Purpose and Utility

This concept presents means to increase the effectiveness of planetary "gravity turns" around solar system bodies with an atmosphere by using downward aerodynamic lift forces to more sharply bend the spacecraft's trajectory around a planetary body than is possible using gravity forces alone. Its implementation would result in much shorter, less expensive planetary missions.

16.7.2 Principles of Operation

Using the gravity field of major planets to alter the velocity vector of a planetary spacecraft is widely employed and greatly reduces the energy requirements for many planetary exploration missions. These "gravitational collisions" are used to impart more velocity to the spacecraft and turn its velocity vector in inertial space, limited by how close to the planet's surface the approach can be dared and the strength of the planet's gravity field. The system is illustrated in Fig. 16.7.

Downward aerodynamic lift can be used around bodies with an atmosphere to bend the trajectory in a smaller radius of curvature if the spacecraft is encased in an appropriate hypersonic aerodynamically shaped shell and the lift vector when skimming a planet's atmosphere is directed toward (rather than the usual direction away from) the planet. This combined aero-gravitational planetary turn can have large advantages compared with using gravity alone. An added advantage is that some planetary missions using this technique can dispense with a close pass by Jupiter, avoiding the intense radiation and its effects on spacecraft electronics.

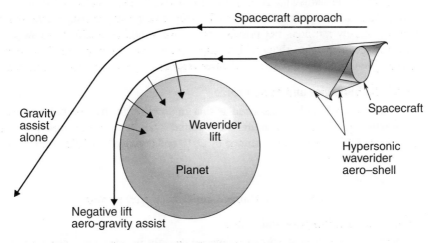

Performance Improvements (km/sec)			
Assist Type	Jupiter	Mars	Venus
Gravity	15	3–5	3–5
Aero-gravity	–	30	15

Faster Missions		
Solar System Body	Gravity	Aero-gravity
Sun	5 years (VEJS)	4 months (EVS)
Pluto	15 years (EJP)	5 years (EVMP)

Fig. 16.7. Aero-gravity assist for interplanetary spacecraft.

16.7.3 Characteristics of the Illustrated System

The system encases the spacecraft in a hypersonic wave rider or other form of aerodynamic shell, which could be jettisoned after the maneuver or kept for repeated encounters. A specific design was not attempted in this study, though it is estimated that typical vehicle lift/drag ratios of about 15 could be attained. The entry Mach number would be about 25–50, and both the spacecraft and the aeroshell would have to resist loads of 10–20 g.

16.7.4 Performance and Weight

Typical of what could be attained using this combination technique is almost an order of magnitude increase in the effective velocity change over gravity alone, with increases from 3 to 30 kilometers per second at Mars and half that at Venus. Another measure is the time required to reach a planetary target using this type of maneuver because for some missions enough velocity can be gained without passing close to Jupiter. With such effectiveness from using Venus, Mars, or Earth for turns, close flybys of Jupiter can be avoided, thus reducing the radiation dose the spacecraft has to withstand. A solar mission would be reduced from 5 years to 4, and a Pluto mission from 15 years to 5, using aero-gravity assist. In addition, a number of new missions not possible otherwise would be enabled, including a Saturn magnetospheric orbiter with multiple plane changing flybys of Titan; plane changing multiple Earth passes; and a free return interplanetary shuttle mission.

16.7.5 Technologies and Time Frame

The only new technology required is that of lightweight hypersonic waverider aeroshells capable of withstanding the required loads at elevated temperatures. It is estimated this can be demonstrated by 2010 and deployed by 2015, making this a midterm concept.

16.8 Global Range Gun/Orbital Accelerator

16.8.1 Concept Purpose and Utility

With this concept a payload can be placed directly in orbit as a "one stage gun launcher" or used as artillery to deliver a shell or payload anywhere on the globe with precision at near orbital speeds and with artillery-like operations complexity and cost. Its implementation would result in systems with great operational utility, such as precision strike weapons with global reach or lower-cost launch vehicles to orbit smaller payloads alone or many packages to be assembled in orbit.

16.8.2 Principles of Operation

The concept, illustrated in Fig. 16.8, is a blast wave accelerator. A series of hollow explosive rings are detonated in sequence, causing a near-constant pressure at the base of the payload projectile and thus a near-constant and large acceleration. Mach 27 or more can be attained in a short gun. The gun can have a barrel, or the

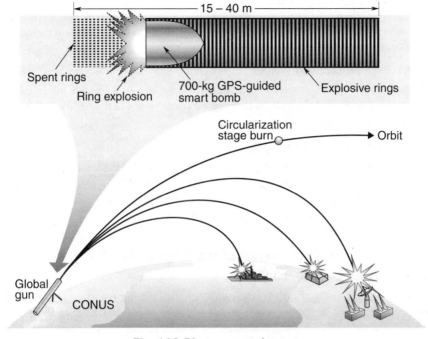

Fig. 16.8. Blast wave accelerator.

rings can be supported by a top beam as the explosions are so fast that inertial confinement may suffice.

The theory is of recent Russian origin and verified in a study funded by NASA. The gun can accelerate a payload to near-orbital velocity, with the payload travelling a ballistic path to anywhere in the world. Alternatively the payload can have an apogee kick motor to circularize the trajectory and enter orbit. The structure is simple, with replacement of explosive rings and rudimentary structure being the principal change to reload the gun, which may result in nominal-to-low costs, though the high accelerations of 100,000 to 300,000 g will limit payload types and demand constructions that can survive them.

16.8.3 Characteristics of the Illustrated System

Mach 27 exit velocities result from use of a 15-meter long barrel and an acceleration of 300,000 g, and a 40-meter long barrel would require only 100,000 g. A payload could be a 700-kilogram GPS-guided glide bomb or a payload for space application. The concept could be useful for launching small dense payloads or multiple modules designed to be assembled in space. In addition to launching commodities such as water, propellants, and raw materials, it could launch properly constructed smaller spacecraft containing electronics. (Remember that in World War II, variable-time fuses withstood 75,000 g using vacuum tube electronics.)

These blast-wave accelerators could be situated in CONUS for safe global artillery. They could be used to throw oriented projectiles or explosives against surface and subsurface facilities as well as homing torpedoes that use supercavitation to cruise supersonically underwater.

16.8.4 Performance and Weight

Estimates vary on the cost of a gun launch from a low of $200 per kilogram to more than $2000 per kilogram, depending on construction and whether refurbishment is a viable option.

16.8.5 Technologies and Time Frame

The explosive rings are state of the art, as are systems for their timed sequential explosions. Considerable dynamics analyses are required for a particular design. A concept demonstration could probably take place before 2010, and an operating system deployed before 2015, making this a midterm concept.

16.9 Electromagnetic Launch Assist (Maglifter)

16.9.1 Concept Purpose and Utility

This reusable first stage Earth-bound catapult for increasing the payload mass and reducing the costs of space launch would result in reductions of launch cost to orbit by factors of 2 to 3 without making substantial changes to the launch vehicle itself. It is applicable to many types of launch vehicles.

16.9.2 Principles of Operation

The concept, illustrated in Fig. 16.9, consists of placing a long track along an incline on or inside a mountain with exit at an angle. A sled on the track can be accelerated and then decelerated after it releases its payload and returned empty to the starting point. A magnetically levitated sled is used, accelerated by a linear electric motor on the track. The track has a run-out section that permits smooth deceleration of the sled, allowing its reuse many times. The catapult uses only electricity to accelerate the sled and liquid nitrogen to cool the superconductive magnets levitating the sled above the track. The velocity added by the catapult increases a mounted launch vehicle's normal payload capability. This velocity increase is greatest when the exit angle is about 45 degrees and when the altitude at the track exit from the tunnel is at or above 3000 meters to minimize the effects of atmospheric drag.

16.9.3 Characteristics of the Illustrated System

The system illustrated in Fig. 16.9 is built inside a mountain, for example, in Colorado. It consists of a guideway, tunnel enclosure, and power supply for the electromagnets that levitate and propel the sled. The vehicle exits the track at about Mach 1. The altitude of the track exit is 3000 meters, and its angle to the horizontal is about 45 degrees.

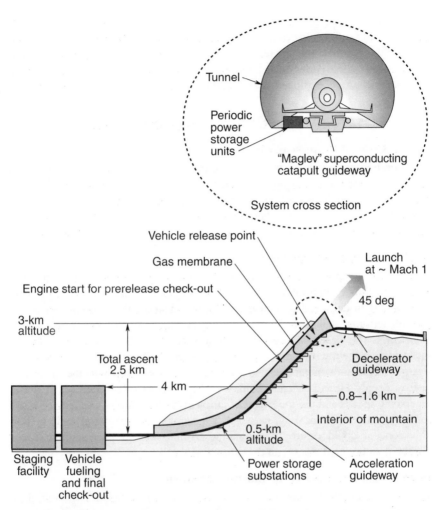

Fig. 16.9. Very low cost "first stage" catapult for ETO (Earth to orbit) launch.

16.9.4 Performance and Weight

The effect of the sled's exit conditions double or even triple the vehicle's normal payload, depending on whether the vehicle is single or two stage to orbit and reusable or expendable. The launch track uses only electricity whose costs are less than $2 per kilogram of payload, and the sled is fully reusable, being brought to a halt after separating the vehicle and returned to the starting point.

16.9.5 Technologies and Time Frame

The technologies for levitating and accelerating a sled are the same as those being developed for high-speed trains. The rocket launch vehicle requires integration with the sled and means for separation without rebound and possible contact with the sled at transonic velocities. A disposable membrane may be needed to maintain tunnel vacuum. While the technologies are not demanding as this is a ground-based system, the development and demonstration of a system could take until 2010 and the system deployed in 2015, probably making this a midterm concept.

16.10 Hard and Submerged Target Kill from Space

16.10.1 Concept Purpose and Utility

Precision non-nuclear force projection to ground, underground, sea, or undersea targets is attained from orbiting spacecraft. Concept implementation would result in a capability to deliver weapons globally with pinpoint accuracy and little collateral damage, without warning, and without placing air crews in harm's way. This concept, illustrated in Fig. 16.10, does not use weapons of mass destruction.

16.10.2 Principles of Operation

Non-nuclear weapons such as kinetic kill devices can be placed in orbit when needed or stationed in orbit for later use. These weapons can be deorbited on command and then delivered to their desired targets using GPS, homing seekers, or both for terminal guidance. They could also be placed in ballistic or boost-glide trajectories. The weapons, singly or in clusters, can be protected during reentry by encapsulating them with thermal protection materials. They can be designed with a large length-to-diameter ratio so that they penetrate deep into the ground to destroy deeply buried concrete bunkers. They can even penetrate sea water at supersonic speeds and use the supercavitation effect to reduce water drag to levels comparable to that of air drag, so that they can cruise underwater at sonic speeds. This concept could destroy a submarine by non-nuclear means—high speed impact of an oriented mass with the submarine's hull.

16.10.3 Characteristics of the Illustrated System

The illustrated system is notional, and no design was undertaken. However, similar systems investigated by government laboratories indicate feasibility and practicality for a number of these applications.

16.10.4 Performance and Weight

The cost of such delivery can be less than that of traditional air-delivery means despite the usual high costs of access to space and space hardware, when infrastructure and operation costs are properly accounted for. Thus, in the future as launch and spacecraft hardware costs drop as described in this book, weapon delivery from space could well become the least expensive way to accomplish a particular objective. In addition, delivery of strikes from space provides the additional effect of complete surprise. As a result the number of strikes required to achieve a particular effect can be greatly reduced and air crew losses greatly reduced.

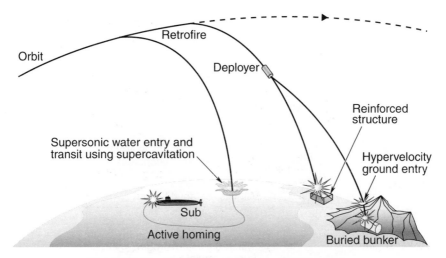

Fig. 16.10. Hard and submerged target kill from space.

16.10.5 Technologies and Time Frame

Technologies of launch and reentry are in hand. Supercavitation effects are reportedly understood. Guidance of precision munitions from space using GPS should be no different from air delivery but needs demonstration in an integrated attack system, both for ground and sea targets. Thus the system could probably be demonstrated from space by 2010 and deployed by 2015, making it a midterm concept.

16.11 Combination Propulsion System Single Stage to Orbit

16.11.1 Concept Purpose and Utility

Launch costs can be reduced while more than doubling payload capability through synergistic use of several technologies previously advocated separately. Implementation will lower cost of access, which will have dramatic impact on use of space. The system is shown in Fig. 16.11.

16.11.2 Principles of Operation

The approach takes advantage of the higher specific impulse of air-breathing engines as compared with rocket engines early in the trajectory, where their specific impulse is greatest. It switches over to rocket propulsion at Mach 6, a much lower number than did the National Aerospace Plane Program, which attempted to make the switch at Mach 18. This forced that program to cruise for long times in the high atmosphere, which created so much drag that the increased hydrogen fuel requirement and heating offset the specific impulse advantages of air-breathing engines.

The technique here uses air-breathing engines as accelerators in high-angle trajectories rather than in a cruise mode, and switches over to rocket propulsion at Mach 6–8. An additional feature is the launch of such a vehicle on a magnetically

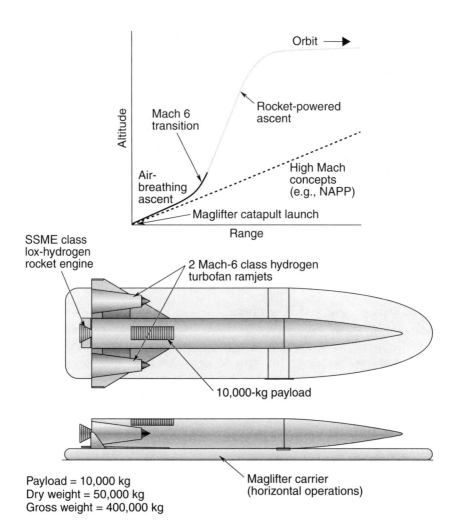

Fig. 16.11. Combination propulsion system SSTO: Maglifter, Mach-6 airbreather, rocket accelerator.

levitated Maglifter sled, which doubles the launch vehicle's payload by itself. These advantages are extraordinarily difficult to prove in design studies and computer simulations, as practically all the phenomena to be modeled are nonlinear and interact. Therefore a flight program will ultimately be required to demonstrate the viability and practicality of this concept.

16.11.3 Characteristics of the Illustrated System

The lower thrust to weight ratio of air-breathing engines compared with rockets makes this a challenging task; however, it could reduce the vehicle dry weight by a factor of 2, the number of rocket engines by a factor of 6, and the total vehicle size considerably, and therefore the acquisition and operations costs of the launch vehicle. In addition the concept eliminates the need for an actively cooled thermal protection system as well as the need for high temperature structures, which were a major stumbling block for the National Aerospace Plane Program. The air-breathing propulsion also provides the system with an inherent capability to ferry itself to and from the launch and landing sites, simplifying the system operations.

16.11.4 Performance and Weight

The example in Fig. 16.11 uses one rocket engine of the Space Shuttle Main Engine class rocket engine, two Mach-6 class hydrogen turbofan engines, and a Maglifter sled catapult. Its dry weight is only 50,000 kilograms for a payload of 10,000 kilograms to a 160-kilometer orbit, easily twice as efficient as a comparable all-rocket SSTO launch vehicle.

16.11.5 Technologies and Time Frame

The technologies of the Mach-6 turbofan engine—its integration into an all composite airframe and thermal protection system—require a development program; however, it is less challenging than that for a single stage to orbit reusable vehicle. Thus the demonstration of all technologies operating together could probably be carried out by approximately 2010, and the system could be deployed in the 2015 time frame making this a midterm concept.

16.12 Fast LEO to GEO Transfer Without Propellants

16.12.1 Concept Purpose and Utility

Payloads can be transferred from LEO to GEO in about the same time as they can be with chemical propulsion, but with the high specific impulse of electric propulsion. The implementation of this concept would result in a much greater payload than conventional chemical stages, yet without incurring the long transfer times, which is the characteristic of low thrust electric propulsion. In addition, when used in a two-way mode, essentially no propellants would be required.

16.12.2 Principles of Operation

Chemical stages have high thrust resulting in LEO to GEO transfer times of 6 hours, but are limited to a specific impulse of about 450 seconds. Electric propulsion stages have specific impulse of 3000–5000 seconds and so require an order of magnitude less propellant, but their low thrust results in transfer times of months.

This concept, shown in Fig. 16.12, uses two heavy rotating reusable platforms in orbit, each having a tether with a smart maneuvering end, which allows the rapid "slinging" of payloads from one platform to the other in Hohman trajectories. Subtraction of tangential from orbital velocity at the lower end of the rotation is used to match approach velocities and catch payloads incoming from lower orbits.

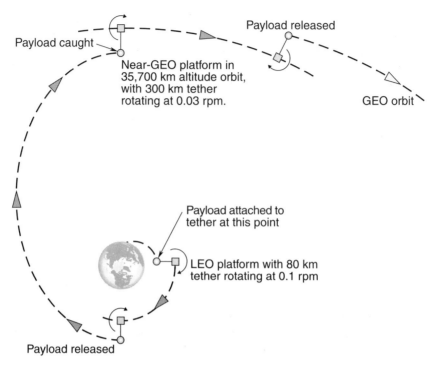

Fig. 16.12. Fast LEO to GEO transfer without propellants: 6-hour transfers at specific impulse of 5000.

Addition of those velocities at the upper end of the rotation and subsequent release is used to sling them to a higher apogee. The energy for the rapid transfers comes from that of the platforms, whose orbits are lowered slightly by the operation. The platforms then reboost to their original altitude using electric propulsion over a longer time. The net effect is an approximately 6-hour transfer from LEO to GEO with an effective specific impulse of 5000, which is not achievable by any other known means.

16.12.3 Characteristics of the Illustrated System

A LEO platform with an 80-kilometer long tether rotating at 0.1 revolutions per minute catches a payload from a suborbital launch vehicle at the lowest point of its rotation. It releases the payload at the upper extreme of its rotation, placing the payload into a Hohman transfer. A second platform in a near-GEO orbit has a 300-kilometer long tether rotating at 0.03 revolutions per minute.

With this design the velocity of the lower end of the tether is matched to the velocity of the approaching payload, which is caught by making any small trajectory corrections needed using a small maneuvering stage at the tether end. The payload is released at the top of the rotation and finds itself in stable GEO orbit.

No propellants need be expended for this transfer, but ion propulsion is required to raise the orbit of the platforms slowly after the transfer. In the future many large spacecraft in GEO will be brought back to LEO for repair or upgrade, and the process of lowering their orbits by using the system in reverse generates energy for the two platforms and raises their orbits. When the average mass traffic up is the same as that down, the system will require essentially no net energy for continuous operation.

16.12.4 Performance and Weight

The total mass of the tethers and platforms breaks even with the mass of chemical stages they replace after a few operations. After that it becomes permanent infrastructure, which can be used by most LEO to GEO transferring spacecraft, saving more with every use. Some increase in the strength of today's tether materials is needed, or alternatively the upper platform must be in elliptical orbit. (If implemented using Buckytubes, the tether weight would be greatly reduced in the far term. Then circular platform orbits can be used, and the mass investment will pay for itself with the first use.)

16.12.5 Technologies and Time Frame

The technologies to be developed are those of long-life multiline tethers and small maneuvering stages with capture hardware. Full demonstration of the dynamic operations will be required. The technologies and operations could probably all be demonstrated by 2010 and deployed by 2015, making this a midterm concept.

16.13 Earth–Moon–Earth Transportation System Using No Propellants

16.13.1 Concept Purpose and Utility

This concept presents a means to travel from Earth orbit to the moon's surface and back without requiring propellants. Its implementation would establish a permanent infrastructure whose low cost of use would facilitate future lunar activities. In principle the concept is applicable to transfer between any two bodies in the solar system. A notional design of the concept is shown in Fig. 16.13.

16.13.2 Principles of Operation

Exploiting the dynamics of tethered rotating platform systems to transfer payloads without propulsion was described in the previous concept for transfers between LEO and GEO. The same principle can be used for transfers between two rotating platforms, one in LEO and the other in low lunar orbit. The platform in lunar orbit can furthermore have an orbit, tether length, and rotation period chosen such that its end touches down on the lunar surface periodically at zero relative velocity. It can then drop off payloads from Earth orbit without expending any propellants.

The system operates as well in reverse by returning payloads picked up from the lunar surface by the lunar orbit platform and released from its tether to be caught by

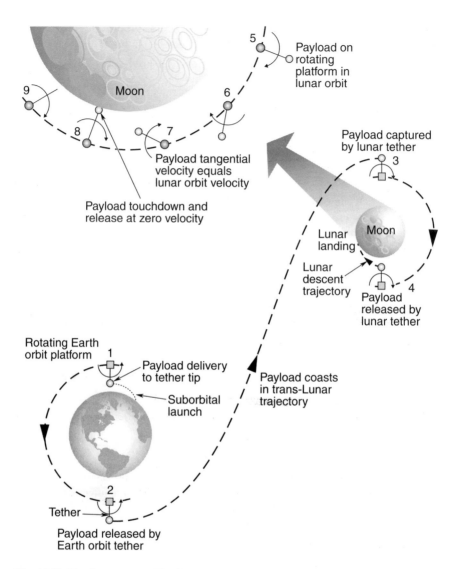

Fig. 16.13. Earth to moon to Earth transportation system requiring no propellants.

the LEO tethered platform. The platforms can initially have high specific impulse electric propulsion to regain their previous orbits using minimal propellants. In the long-term future, when the mass of payloads over time going to the moon may equal that of returning mining or processed payloads, the system would operate with no net energy or propellants (except for making up small losses) as the average energy from the returns would supply that needed for the ascents.

16.13.3 Characteristics of the Illustrated System

This concept was only notionally designed using tether materials with twice the strength of today's fibers, though a similar NASA design feasibility study was performed by Tethers Unlimited. In that study the use of current materials for the tethers resulted in a requirement to place the Earth orbit platform in elliptical orbit and increased the number of system uses before breakeven could be achieved with chemically propelled transfers. Whereas the illustrated system eliminates all propulsive maneuvers, a number of options exist. An example would be to minimize the length of the rotating lunar orbit platform's tether and place it in an orbit whereby the payload is placed or retrieved directly from the lunar surface using a propulsive stage. There are several other options for implementing this concept.

16.13.4 Performance and Weight

When a factor of 2 increase in the tensile strength of today's tether materials is attained the lower platform can be placed in circular LEO. When Buckytube tethers and structures are developed in the far term the system becomes thoroughly practical and extremely lightweight, and breaks even with chemical propulsion in a single payload transfer, thereafter becoming permanent transportation infrastructure.

16.13.5 Technologies and Time Frame

The principal technologies to be developed are those of long-life, long multiline tethers, small maneuvering stages with capture hardware, and homing guidance and capture algorithms. These could probably be demonstrated by 2010 and deployed by 2015, making this a midterm concept.

16.14 Mars Phobos—Deimos Ladder

16.14.1 Concept Purpose and Utility

A permanent infrastructure is emplaced around Mars and acts as a ladder for propellantless approaches to, and departures from, the planet. Its implementation would greatly reduce spacecraft weight for Mars exploration or science missions and pay for itself in just four uses.

16.14.2 Principles of Operation

A ladder of rotating tethers is placed around Mars for slinging payloads toward and away from the surface without using propellants. The operation is similar to that of the LEO–GEO and the LEO–moon tethers of the previous two concepts, except that nature has already provided the two heavy platforms in Mars orbit— Phobos and Deimos. These moons of Mars are in fortuitous orbits such that fixed tethers extended from them "toward and away from" the planet can be used to impart very large transfer velocities to payloads in the same mode as in the last concepts. The high velocities are attainable for physically realizable tether lengths and masses. Tethers will each be fixed to the moons and stabilized by the gravity gradient in Mars orbit to attain local vertical orientations. Since the Martian moons have rotation periods synchronized with their orbit periods the rotation of the tethers will attain the same slinging effects as provided by more rapid rotation of the tethers in the previous concepts.

An approaching payload from Earth would rendezvous with the upper end of the upper Deimos tether, crawl along the tether to Deimos, transfer to the other tether, crawl along it away from Deimos to the tether end, and let go. Properly designed, the result is a payload Hohman transfer, which will intersect with the tip of the upper tether extending from the lower moon, Phobos. Repeating this operation would find the payload, after release from the lower Phobos tether, in a suborbital trajectory to enter the Mars atmosphere and descend to the surface without using propellants. This system is completely reversible, enabling a suborbital launch from the Martian surface to result in a trans-Earth escape trajectory, also without using propellants.

16.14.3 Characteristics of the Illustrated System

In the system illustrated in Fig. 16.14, the upper Deimos tether is 6100 kilometers long, and the lower tether is 2960 kilometers long. The upper Phobos tether is 1100 kilometers long, while the lower one is 940. Though these tethers are long compared with those contemplated for use in Earth orbit, their mass is reasonable due to the lower gravity field of Mars. These tether lengths are calculated to provide for nominally zero relative velocities between their ends and approaching/departing spacecraft in transfer orbits.

A spacecraft can start in a 375-kilometer Mars orbit or at the surface and inject into transfer orbit and proceed upward through the ladder, leaving the upper Deimos tether with sufficient velocity to transfer directly to Earth without expending any propellants. Likewise, a spacecraft approaching from Earth can rendezvous with, and attach to, that tether and proceed downward through the ladder to arrive in 375-kilometer Mars orbit by aerobraking or enter a Mars surface descent trajectory, all without propellant consumption.

16.14.4 Performance and Weight

The system saves 3830 meters per second propulsive velocity increment when going either up or down the tether ladder. Since this velocity change would require a propellant mass of 3.3 times the mass of the final landing vehicle and 6.6 times the mass of the final Earth-return capsule after ascent from Mars, the ability to avoid carrying these propellants means that the Earth takeoff mass of a Mars mission can be reduced by an order of magnitude, including overhead effects. This, in turn, reduces the number and cost of launches to assemble a Mars spaceship by roughly the same factor, bringing the costs of Mars missions down by an order of magnitude.

The tethers would represent a massive permanent infrastructure, yet they will pay for themselves in just four missions even when using conventional tether materials compared to the cost of using chemical propulsion. (If implemented using

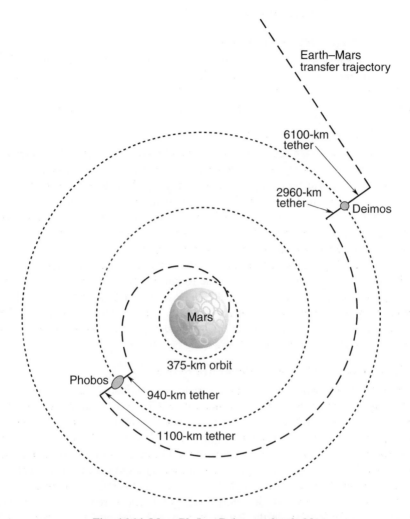

Fig. 16.14. Mars Phobos-Deimos tether ladder.

Buckytubes throughout, their total weight would be reduced in the far term by a factor of 100, making the tether ladder inexpensive and extremely attractive as permanent infrastructure that pays for itself in the first use.)

16.14.5 Technologies and Time Frame

The technologies required are those of large, long, massive, multiline fail-safe tethers; mechanisms for spacecraft to rendezvous with the tether ends, attach themselves, and roll along the tethers winching themselves up or down; and homing/guidance algorithms. These can probably be demonstrated by 2015 and deployed by 2020, making this a far term concept.

16.15 Nuclear Waste Disposal in Space

16.15.1 Concept Purpose and Utility

Hazardous nuclear wastes are disposed of in deep space using this concept, either in a solar orbit or on a designated site on the moon. Implementation of this concept would solve the politically difficult problems of where to store nuclear waste in the United States (or worldwide) because the waste would not be stored on Earth.

16.15.2 Principles of Operation

The concept depends on separating high-level actinides from the bulk of nuclear-reactor waste. These are the most troubling of the waste products because they are very long lived. Fortunately they represent only a small fraction of the total waste products of nuclear fission power plants, and thus the mass to be launched is minimized. Studies done in the 1960s and 1970s indicated that the waste could be packaged in containers so constructed as to withstand any launch vehicle explosion or aborted reentry from space without breaking open.

With this packaging and the increased reliabilities of launch vehicles expected in the long term, it would be conceivable to launch the waste into space, as the consequences of failed launches would be no worse than those of launch vehicles not carrying nuclear waste. The waste could be placed into a long-term stable solar orbit or landed on the moon for long-term storage, yet could be recovered by future generations should they so desire. (See Fig. 16.15.)

16.15.3 Characteristics of the Illustrated System

Requirements for launch vehicle performance are similar to those of Mars or Venus exploration missions. Waste would not be delivered to sun-impact trajectories to assure burn up because demands on launch velocity would be too great. Waste packaging was designed as multiple concentric spheres of beryllium so constructed to prevent rupture even in cases of total vehicle explosion on the launchpad, full velocity reentry from space and impact on a granite mountain peak, or descent into the deepest ocean trough. In each scenario the integrity of the package could be guaranteed for decades until it was safely recovered. Constructing the spheres from Buckytubes in the longer term would guarantee their integrity far longer, through greater accidents and for far lower added packaging weight.

The reliability of launch vehicles in the far term future will be as great as that of modern airliners, with catastrophic accident rates under one per million launches, and once in space their stable location can be guaranteed for millions of years. Given that the integrity of the waste packages need only be guaranteed for decades until recovery, the risks of long-term nuclear waste storage in space are comparable to or lower than that of their storage anywhere on Earth.

16.15.4 Performance and Weight

Calculations for a reference system indicated that even at current launch costs of $12,000 per kilogram, the added costs of the packaging and destination propulsion,

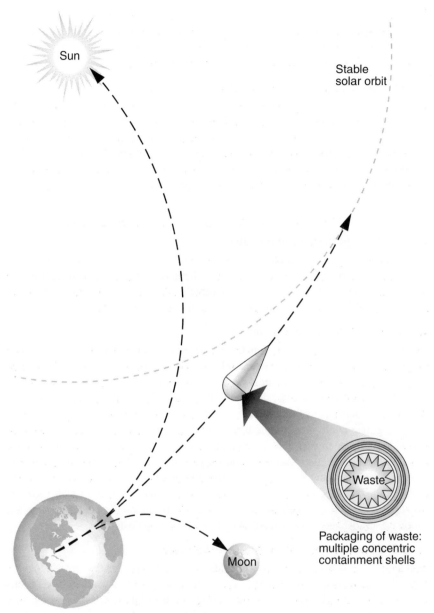

Fig. 16.15. Nuclear waste disposal in space.

when converted to costs added to electricity generated by nuclear power, resulted in less than 2 cents per kilowatt-hour additional consumer price. This is only 3 to 30 percent of the typical energy costs today, yet will probably be an order of magnitude lower in the long-term future as the cost of power rises because of shortages and environmental concerns. (If implemented using Buckytubes throughout the launch vehicle and the payload packaging canister, their total weight would be reduced in the far term by a factor of 100, and the above cost fraction would drop to 0.3 to 3 percent even at current prices, a truly negligible cost.)

16.15.5 Technologies and Time Frame

The technologies requiring development are principally those for building unbreakable containers, increasing launch reliability by several orders of magnitude, and implementing upper or transfer stages with high performance yet the same high reliabilities. These technologies have been discussed previously and though difficult could probably be demonstrated in the 2015 time period and deployed in the 2020 time period, making this a far term concept.

16.16 RF-Powered Electromagnetic LEO Launch System

16.16.1 Concept Purpose and Utility

This concept presents a launch vehicle that requires almost no onboard propellants, and promises to reduce costs of space launch by three orders of magnitude from today's levels. It is one of a class of vehicles that, if implemented, would create an industrial revolution in space resulting from the low costs of access.

16.16.2 Principles of Operation

As illustrated in Fig. 16.16, millimeter wave or microwave energy is generated on the ground and beamed to the launch vehicle. There it diffracts around the shaped vehicle and is focused at a point ahead of the vehicle, causing atmospheric breakdown due to the intense energy density at that point. The resulting ionized air flows downward around the vehicle and is accelerated by superconducting coils at its base using magnetohydrodynamic forces. This accelerated air produces downward thrust akin to a jet engine. Since the reaction mass is free air, there are no onboard propellants, and the energy source is on the ground, the vehicle specific impulse is infinite. Therefore launch energy costs stem principally from the costs of electricity to power the beaming system. Nonetheless a small internal source of hydrogen and oxygen is used as the propellant in a rocket mode for the final ascent to orbit when out of the atmosphere, somewhat adding to its operating costs.

16.16.3 Characteristics of the Illustrated System

Accelerated air is ejected through nozzles at the periphery of the vehicle's base. Multiple superconducting coils are used in groups around the periphery of a lenticular-shaped vehicle, which allow hovering and translation sideways much as a helicopter. The vehicle's coils can act together to propel the vehicle toward orbit.

The ground system comprises a power supply, high power microwave or millimeter wave beam, and a phased array antenna to form and propagate a narrow beam toward the spacecraft. Automated tracking and safety systems would ensure controlled energy flow to the vehicle. The system is completely reusable and consumes no propellants except for the hydrogen and oxygen used for final ascent to orbit.

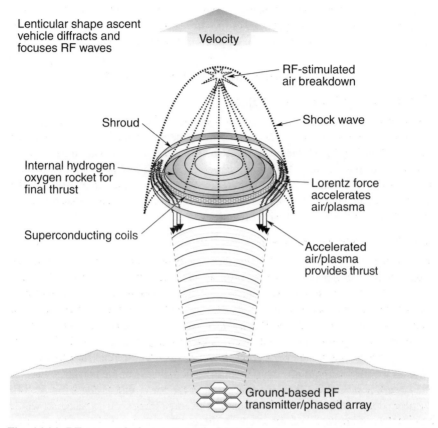

Lenticular shape ascent vehicle diffracts and focuses RF waves

Velocity

RF-stimulated air breakdown

Shroud

Shock wave

Internal hydrogen oxygen rocket for final thrust

Lorentz force accelerates air/plasma

Superconducting coils

Accelerated air/plasma provides thrust

Ground-based RF transmitter/phased array

Fig. 16.16. RF-powered electromagnetic LEO launch system. Most launch vehicle energy is supplied externally.

A laser-powered system in which a ground laser caused air breakdown and propelled a passive aerodynamic shape upward has already been tested at small scales and low velocities, demonstrating some of the required principles.

16.16.4 Performance and Weight

This system has been proposed and analyzed by Leik Myrabo of Rensselaer Polytechnic and supported by NASA. It promises to eventually result in launch costs of $20 per kilogram to orbit. (If implemented using Buckytubes throughout, in the far term its dry weight would be reduced by a factor of 100, and its operating costs by a factor of 10 to about $2 per kilogram.)

16.16.5 Technologies and Time Frame

A number of technologies require development, including the high energy and power-beaming millimeter or microwave system, the superconducting coils for the vehicle and their controls, and the vehicle design to satisfy aerodynamic, chemical rocket, and magnetohydrodynamic requirements. These technologies could probably be demonstrated acting together in the 2020 time frame and a vehicle deployed in the 2025 time frame, making this a far-term concept.

16.17 Geostationary LEO/MEO Spacecraft

16.17.1 Concept Purpose and Utility

A stationary spacecraft is described that hovers over the Earth exactly in the manner of a geostationary spacecraft, but at low or medium altitudes instead of in the GEO orbit at 36,000 kilometers altitude. Its implementation would allow a single spacecraft to provide regional or theater coverage from low or medium altitudes, as opposed to the many satellites required in free orbits at the same altitudes. The spacecraft can be used for civilian, commercial, or defense functions such as communications, sensing, or military force projection.

16.17.2 Principles of Operation

In the system shown in Fig. 16.17 a tether is suspended from a super-geostationary countermass and extends to a low altitude of choice. If conventional materials were used to construct the tether, its mass would be hundreds of thousands times greater than the payload it was designed to suspend, making it impractical. But a tether constructed entirely of Buckytubes would have a much smaller mass, approximately equal to twice that of the suspended payload spacecraft. The countermass, which could be simply a collection of dead GEO satellites, could equal the mass of the payload to place the system in static balance. If the center of mass of the system is placed in GEO, the spacecraft at the lower end of the tether will be suspended and stationary over the equator. Coverage from such a spacecraft would be circular, stationary, and centered on the equator. If the center of the system mass is placed into inclined synchronous orbit, the coverage is the same diameter but moves along a "figure 8" ground track centered on the equator, with extreme latitudes equal to the orbit inclination.

16.17.3 Characteristics of the Illustrated System

The illustrated system shows a geostationary low altitude spacecraft in which the tether is about 72,000 kilometers long from the spacecraft to the countermass, with the center of gravity being in GEO. The length of the tether upward depends on the countermass available as well as on the tether mass. Once installed, this system becomes permanent launch infrastructure that can be used by all payloads travelling between GEO and Earth.

16.17.4 Performance and Weight

When the system is constructed entirely from pure Buckytube materials the system mass (weight) will be only 3 to 4 times greater than that of the design payload mass. Earth visibility from the payload will be at least 2700 kilometers from 160-kilometer altitude, increasing rapidly with suspended payload altitude increases to at

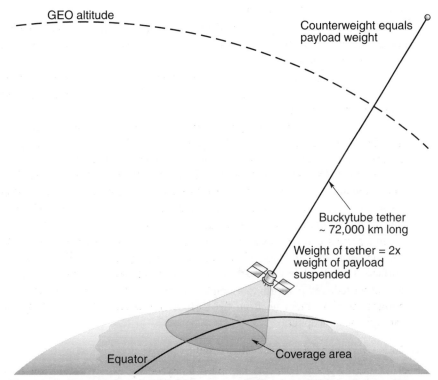

Fig. 16.17. Geostationary LEO/MEO spacecraft. Earth visibility increases with suspended payload altitude increases; i.e., for an orbit altitude of 160 km, instantaneous coverage is 2700 km; for 320 km, 3800 km; 960 km, 6500; and 3200 km, 10,700 km.

least 10,000 kilometers from 3,200-kilometer altitude. The number of spacecraft that can be so suspended is essentially limited only by considerations of collision avoidance for ordinary spacecraft and their launches.

16.17.5 Technologies and Time Frame

The development and demonstration of Buckytubes of the length and redundancy required for reliable functioning could probably occur by the 2020 time frame, and their deployment in this system configuration in the 2025 time frame, making this a far term concept.

16.18 Macroengineering Space Concepts

16.18.1 Concept Purpose and Utility

A variety of large structure approaches are presented to create a permanent, massive Earth-to-orbit launch infrastructure. Implementation of any would greatly reduce costs of access to space.

16.18.2 Principles of Operation

The systems illustrated in Fig. 16.18 are macroengineering projects comparable to major constructions on Earth, such as the largest bridges and dams. They are (from the left in the illustration)

- a very long magnetically levitated catapult that extends above most of the atmosphere and accelerates payloads at low g directly to orbital velocities above most of the atmosphere
- a circulating stream of ferromagnetic pellets that is magnetically accelerated and shot into space, turned around magnetically in GEO, turned around again magnetically underground, and recirculated through an atmospheric shield in a vacuum tube. The magnetic drag of the high-velocity pellet stream causes levitation and acceleration of a simple one-turn elevator, causing it to rise to GEO
- a smaller version of the pellet stream levitator that transfers the payloads to a low orbit platform from a tether balanced about GEO
- a similar tether that reaches all the way from the surface to further than GEO for counterbalance and has a simple elevator for carrying payloads from the surface to GEO or beyond, where they would simply let go to inject
- a rotating tethered platform that emplaces payloads directly from LEO on the surface, or picks them up from the surface and delivers them to LEO, both through the atmosphere

16.18.3 Characteristics of the Illustrated Systems

Each of these system concepts has been analyzed by experts, and their basic physics hold up. However, they are such massive and expensive undertakings that their realization is probably further away than the 30-year horizon of this study (except for the Buckytube elevator, which may be just at that horizon, and was addressed in the previous concept). The mass to and from space must be several orders of magnitude greater than today for any of these macroengineering programs to be worth the likely investment. Even though such traffic levels are inevitable, they will probably not materialize until at least the 30-year time frame. Nonetheless they have the intrinsic value that none require propellants or chemical engines, all are completely reusable, most work as well in raising payloads as in lowering them, and all have low costs of operation once their investment costs are amortized, much as most terrestrial infrastructures. Because these systems are permanent, active measures would be required to prevent collisions with orbiting objects. These macroengineering concepts are included to give perspective to the other advanced transportation concepts in this report, which may not appear so "wild" in comparison.

16.18.4 Performance and Weight

Preliminary weight and performance numbers exist for each of these concepts, but are not included because of the remote likelihood that any could be implemented in the time horizon of this study. The advent of Buckytubes would reduce the weight

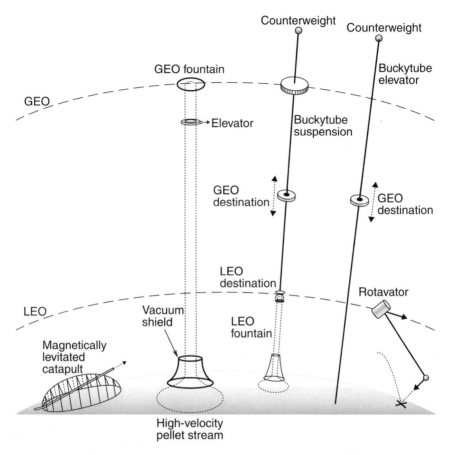

Fig. 16.18. Macroengineering space concepts: Earth-to-space transportation without propellants.

of all these macrostructures by a factor of 100, which should make them more practical and less expensive, but probably still beyond the horizon of this study because of the limited availability of oriented Buckytubes in the massive quantities needed are not expected much before 2025.

16.18.5 Technologies and Time Frame

Extensive structural, assembly, and control technologies make demonstration unlikely before 2025, with deployment probably substantially beyond that date. These are, therefore, far term concepts.

17 Power and Energy Beaming Concepts

These concepts address the beaming of energy or power through space at levels generally much greater than those treated in Chapter 15. Many capitalize on the technologies and techniques of Part II to enable space devices that are more efficient; have higher power, larger aperture, or both; and can be rapidly and accurately pointed to place the power and energy on a distant target or receiver. These concepts are intended for defensive or offensive weapons, as well as for use in remote power delivery for civilian applications.

The 19 concepts in this category are listed in Table 17.1. The organization is the same as in the preceding chapters, with the concepts identified as near term, midterm, far term, and "U" if classified, and presented in rough order of increasing complexity, longer time frame, or both. The concepts contain both radio frequency (RF) and optical beamed energy applications.

Table 17.1. Power Beaming and Energy Delivery Concepts

Time Frame	Concept
Near term	Beam-powered orbit transfer vehicle
	Off-board auxiliary power system for spacecraft
Midterm	Mid term orbital jammer and spoofer
	Passive power distribution via GEO mirrors
Far term	Tunable, ground powered orbital jammer
	Defense uses of COTS (commercial off the shelf) power unit
	Wireless energy distribution satellite
	High power microwave electronics kill (U)
	LEO/MEO solar power satellite
	Provision of energy to lunar bases
	Active Denial fence generator (U)
	Active Denial weapon system (U)
	GEO solar power satellite
	RF-powered electromagnetic LEO launch system
	GEO mirror ground laser force projection
	Ultrafast planetary transport
	Geostationary LEO/MEO power relay using Buckytubes
	Aircraft powering via laser beams
	Interstellar probe experiment

17.1 Beam-Powered Orbit Transfer Vehicle

17.1.1 Concept Purpose and Utility

An efficient orbit transfer vehicle is created that is particularly well suited for LEO to GEO and Lunar missions. Its use could greatly reduce the costs and time for missions going to higher energy Earth orbits and Lunar destinations.

17.1.2 Principles of Operation

The vehicle uses ion or pulsed plasma propulsion and obtains its power from a photovoltaic array illuminated by a laser from the ground. The area of array needed for a given power output is greatly reduced from that of a solar powered vehicle because the laser is able to place a much larger energy density on the array than that provided by the sun. Furthermore the laser wavelength and the array bandgap energy can be chosen so that the array operates at maximum conversion efficiency with the illuminating light. This conversion efficiency is typically 65 percent maximum. The absence of any other portions of the spectrum reduces the heat input and thus the temperature of the array. In addition the laser can be pulsed at a rate that is optimal for powering the pulsed plasma thrusters, thus minimizing the complexity and weight of the power conditioning needed.

17.1.3 Characteristics of the Illustrated System

The illustrated concept in Fig. 17.1 uses gallium arsenide solar cells illuminated at a wavelength of 0.85 microns, their bandgap energy wavelength. The laser parameters are chosen so as to place an energy density of four to five suns on the array. The array has to be oriented toward the laser for power to be received, but the accuracy required is minimal because concentration is not used in this example. Its use would further reduce solar array weight and cost roughly in inverse proportion to the concentration factor, at the expense of more accurate pointing broadside toward the laser. The concept illustrates the synergistic effects of using lasers, bandgap matched arrays, and pulsed plasma thrusters that work together, each at its optimum.

17.1.4 Performance and Weight

The orbit transfer vehicle created with this concept has 25 percent of the weight and 50 percent of the cost of a solar electric vehicle for the same mission, and can transfer from LEO to GEO in 30 percent of the time. The vehicle advantages are even greater for LEO to moon missions, but those advantages have not been quantified.

17.1.5 Technologies and Time Frame

The technologies are in hand. A system demonstration probably could be undertaken by 2005 and the system deployed in 2010, making this a near term concept.

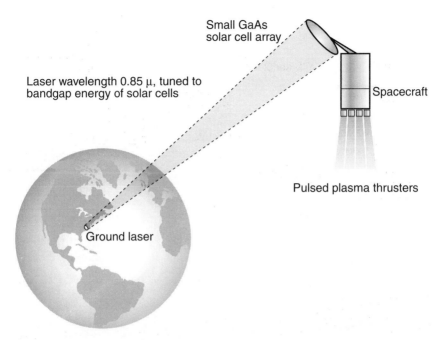

Fig. 17.1. Beam-powered orbit transfer vehicle.

17.2 Off-Board Auxiliary Power System for Spacecraft

17.2.1 Concept Purpose and Utility

This concept presents means for augmenting the electrical power at a separated spacecraft on a commercial basis—an "orbital power and light company." The International Space Station is an example of a user facility in space. Its implementation would enable commercial ventures to provide such power at lower prices than add-on attached power on a government facility.

17.2.2 Principles of Operation

Solar energy is collected in an array on a stand-alone spacecraft and converted to a microwave or millimeter wave beam by a phased array antenna. The beam created by the transmitting array illuminates a rectenna on the International Space Station and creates a direct current output, which is sent to the space station power bus. A reversible electromagnetic tether is placed on the space station and is used as an orbital accumulator instead of batteries for storing the added power. It stores a portion of the power received by raising orbital altitude, which is recovered as electric power when needed while reducing the orbital altitude. The tether can function as a reversible electromagnetic motor generator to store energy, make up drag, maneuver the space station, and provide bursts of higher power when needed. The system is illustrated in Fig. 17.2.

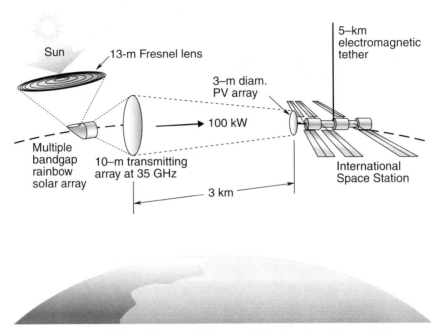

Fig. 17.2. Commercial 50-kW power add-on to the International Space Station as an example of off-board auxiliary power for spacecraft.

17.2.3 Characteristics of the Illustrated System

The illustrated system uses a multiple bandgap, spectrally split array and thin film Fresnel lens concentrator to develop 200 kilowatts, which are then conditioned and fed to a 35-gigahertz transmitter. The transmitter, in turn, feeds a phased array antenna that radiates a 100-kilowatt beam toward the rectenna receiving array on the station, which operates at 85 percent efficiency and feeds a power conditioner/ inverter that supplies at least 50 kilowatts to the space station power bus. The power can be used as received or stored by raising station altitude via the 5-kilo- meter electromagnetic tether. A 10-meter transmitting array is used, which fully illuminates the 3-meter rectenna a distance of 3 kilometers between the power spacecraft and the space station.

17.2.4 Performance and Weight

When sunlit, the stand-alone power spacecraft supplies at least 50 kilowatts full time to the space station, which stores a portion of that power for use during night- time by using the tether as an orbital accumulator. The power unit can be moved to the vicinity of a different space facility and operated as a commercial power deliv- ery service. It weighs only 650 kilograms. (If implemented using Buckytubes throughout, its total weight would be reduced in the far term to about 7 kilograms.)

17.2.5 Technologies and Time Frame

Principal technologies required are the spectrally split, bandgap matched solar array, the Fresnel thin film concentrator, the 10-meter-diameter Ka band antenna array, the 100-kilowatt Ka band transmitter, and the electrodynamic tether. Of these the pacing item is probably the transmitter/antenna combination, which can probably be demonstrated by 2005 and deployed by 2010, making this a near term concept.

17.3 Mid Term Orbital Jammer and Spoofer

17.3.1 Concept Purpose and Utility

A powerful jammer in space is described that can jam most surface and airborne electronics, radars, and communication devices in a theater. Its implementation would be a major force multiplier in tactical situations, and would avoid putting electronic warfare air crews in harm's way.

17.3.2 Principles of Operation

The spacecraft, illustrated in Fig. 17.3, consists of a large membrane RF reflector to create a narrow beam focused on ground or air targets. The feed is a stationkept spacecraft with solar array, high power RF transmitter, and phased array feed, which illuminates the reflector. The reflector is a piezoelectric membrane adjusted by an electron beam in response to an optical figure sensor. The electron beam

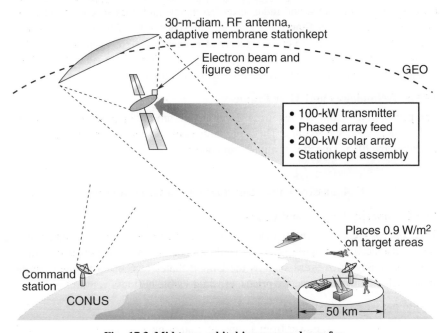

Fig. 17.3. Mid term orbital jammer and spoofer.

generator and the optical figure sensors are both collocated on the feed spacecraft. The power aperture product is so large, particularly at high microwave frequencies, that the jamming signal completely swamps the usual signals present in the receiver front ends of typical electronics.

17.3.3 Characteristics of the Illustrated System

The system uses a 100-kilowatt transmitter and a phased array feed operating at X band. The prime power is a 200-kilowatt solar array. The reflector is a 30-meter-diameter, adaptive piezoelectric membrane shaped by an electron beam. The beam is pointed by positioning the feed spacecraft with respect to the reflector by using MEMS FEEP thruster arrays, as well by the phased array feed. The feed spacecraft translates as well as rotates, whereas the reflector only changes and maintains its attitude in order to minimize the disturbances to the membrane. The reflector has no inflatable torus and is not tensioned, but simply shaped by the electron beam to match the figure reference in the figure sensor.

17.3.4 Performance and Weight

At 10 gigahertz the system generates 10 gigawatts of effective radiated power. This places 0.9 watts per square meter or a 0.2-volt per meter field on the ground from GEO. This power density results in a jamming power of 0.8 milliwatts (−31 decibel watt) in a typical receiver front end. This power is so large compared with normal receiver front end signals that the spacecraft effectively jams most electronics, communications, and radar at its frequency, and does so continuously or as long as necessary in a tactical situation. The spacecraft reflector weight is less than 75 kilograms, and the entire spacecraft probably weighs around 1000 kilograms in orbit. (If implemented using Buckytubes throughout, its total weight would be reduced in the far term to about 10 kilograms.)

17.3.5 Technologies and Time Frame

The principal technologies requiring development are the high power transmitter and the large adaptive antenna. These probably could be demonstrated in 2010 and deployed by 2015. Thus this is a midterm concept.

17.4 Passive Power Distribution Via GEO Mirrors

17.4.1 Concept Purpose and Utility

This concept presents wireless means to transmit electrical power over intercontinental distances. Its implementation would enable commercial power transmission from energy-rich areas such as the United States or Canada to energy-poor areas such as Japan at attractive prices, creating viable business ventures.

17.4.2 Principles of Operation

As shown in Fig. 17.4, power from a conventional or nuclear source on the ground is converted to millimeter waves in a pulsed transmitter and fed into a rectangular

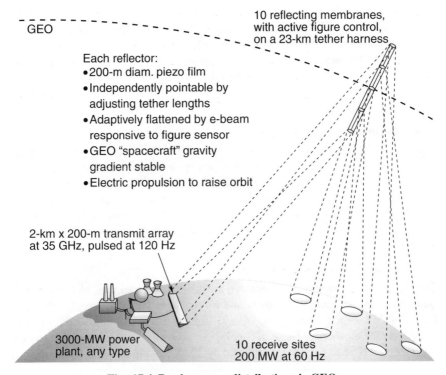

Fig. 17.4. Passive power distribution via GEO.

antenna array. The resulting rectangular fan beam illuminates an array of simple planar reflectors in GEO. These reflectors consist of flat adaptive piezoelectrically shaped membranes whose location and attitude are set and adjusted by means of a series of tethers interconnecting them. The attitude of each reflector is adjusted independently, coarsely via the tethers and finely by MEMS FEEP thrusters, so as to reflect a portion of the incoming millimeter wave energy to illuminate a particular receiving rectenna array on the ground. The length of the tethers is such that the reflectors appear as a projected contiguous set of ellipses just filling the illuminating transmitter fan beam.

Each rectenna receives the energy from one reflector, generates a direct current output, inverts its polarity every other pulse, and thus provides 50–60 hertz electrical power to its users. A 35-gigahertz frequency is used to keep the reflectors, antennas, and rectennas small. A small pumped water storage hydroelectric unit is provided at each receiving site to provide for the infrequent transmission outages caused by heavy rain attenuation.

17.4.3 Characteristics of the Illustrated System

The spacecraft uses 200-meter-diameter adaptive membranes shaped by electron beams responsive to optical figure sensors, whose figure is flattened to a fraction of a millimeter. The tethers connect 10 membranes in a 23-kilometer long ensemble, which is nominally gravity gradient stabilized. Each membrane is finely controlled in attitude and position by arrays of MEMS FEEP thrusters. The transmitter is a 2-kilometer by 200-meter phased array at 35 gigahertz, pulsed at 120 hertz and transmitting 2000 megawatts of power. Each of the 10 rectennas is 2000 meters in diameter and is a wire mesh stretched between poles suspended above a small reservoir, which provides the pumped water storage as well as acts as a security perimeter.

17.4.4 Performance and Weight

The system requires an investment of $6 billion. The entire spacecraft costs about $140 million launched and in GEO. The power delivered could be bought from Canada, for example, at market prices and sold at much higher yet competitive prices in Japan, with the business venture generating an internal rate of return of 35–55 percent. The "spacecraft" weighs about 25,000 kilograms. (If implemented using Buckytubes throughout, its total weight would be reduced in the far term to about 250 kilograms.)

17.4.5 Technologies and Time Frame

The principal technologies to be matured are adaptive piezoelectric electron beam-shaped membranes, long-life tethers and mechanisms for continuously adjusting them, MEMS FEEP thruster arrays, and high power transmitters and antenna arrays at 35 gigahertz. These technologies probably could be demonstrated together by 2015 and the system deployed by 2020, making this a far term concept.

17.5 Tunable, Ground-Powered Orbital Jammer

17.5.1 Concept Purpose and Utility

The concept presents an extremely powerful jammer in space, which is powered entirely by a ground laser beam. Its implementation would result in a capability to jam essentially any and all electronics, radar, and communications devices in a theater even through their sidelobes, and constitutes an extremely useful and effective electronic warfare capability for tactical use.

17.5.2 Principles of Operation

The spacecraft, shown in Fig. 17.5, consists of a large reflector membrane that creates the narrow beam, which is focused on ground or air targets. The antenna feed is illuminated by a dual wavelength high energy laser from the ground, and the RF power is generated directly by heterodyning the two wavelengths of the laser light, producing the RF as the difference between the two laser wavelengths. The RF therefore can be rapidly changed over an extremely broad frequency range by simply shifting one of the illuminating wavelengths. The feed structure is an integrated direct conversion light to microwaves sandwich and illuminates the antenna reflector, which is a piezoelectric membrane adjusted by an electron beam in response

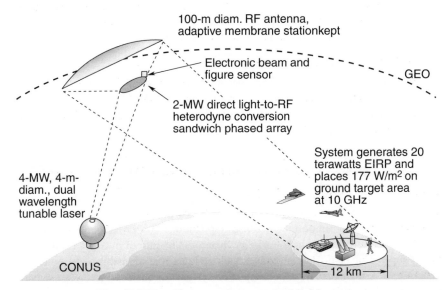

Fig. 17.5. Tunable, ground-powered orbital jammer.

response to an optical figure sensor. The spacecraft has neither a transmitter nor a conventional phased array transmitting antenna.

17.5.3 Characteristics of the Illustrated System

The system uses a 4-megawatt, 4-meter-diameter laser on the ground radiating on two variable separation wavelengths. The laser energy is received in a 5-meter sandwich that converts it directly into 2-megawatt RF, illuminating the 100-meter-diameter reflector. A space-fed lens design could also be used. The frequency can be changed by simply tuning one of the ground lasers, although the limits of the sandwich light to RF converter have not been explored. Pointing the beam is accomplished by changing the relative formation positions of the feed and the reflector. An option exists to use a phased array feed excited by the RF source, and steering the beam by shifting its phase center.

17.5.4 Performance and Weight

At 10 gigahertz, the system generates 20 terawatts of effective isotropic radiated power (EIRP). This places 177 watts per square meter or 2.6 volts per meter field on the ground from GEO, and results in a jamming power of 0.16 watts (−8 decibel watt) in a typical receiver front end. This power is so large compared with normal receiver signals that the spacecraft effectively jams all electronics, communications, and radar at its frequency even through the sidelobes of their antennas, and so will be effective at any incident angle. It can do so continuously or as long as necessary in a tactical situation. The spacecraft reflector weight is less than 750 kilograms, and the entire spacecraft probably weighs approximately 1000 kilograms in orbit. (If implemented using Buckytubes throughout, its total weight would be reduced in the far term to about 10 kilograms.)

17.5.5 Technologies and Time Frame

The principal technologies requiring development are the high power direct conversion sandwich, the large adaptive antenna, and MEMS FEEP thrusters. These probably could be demonstrated by 2010 at subscale but not until 2015 at full power and deployed by 2020, making this a far term concept.

17.6 Defense Uses of Commercial Off the Shelf (COTS) Power Unit

17.6.1 Concept Purpose and Utility

This concept presents a means to use a high power space module designed for commercial use for important defense applications. Its implementation would result in a powerful jamming capability that could be placed into any orbit, capable of supporting tactical operations anywhere in its area of visibility. It could also function as a high power, onboard power system for many spacecraft used for defense applications.

17.6.2 Principles of Operation

It is very likely that the commercial sector will develop and test a module to demonstrate the feasibility of generating gigawatts of power in space and sending them to Earth via a microwave beam. The likely power level of such a module will be 1–5 megawatts. These tests will be used to evaluate the business feasibility of private ventures to emplace spacecraft to provide larger amounts of power, ranging from an initial 400 megawatt level to an eventual 5000 megawatts, to user communities worldwide. A second build of this demonstration system could be used with a modified or different phased array as an unusually powerful space-based jammer or spoofer to place a large microwave power density on the ground or in the air. The field strengths delivered would be large enough to overwhelm most communication equipment over a large tactical area, and the coverage could be directed at any location in a hemisphere. Alternatively this module could be used as the power source for a host of defense mission spacecraft at a power level of 2 megawatts.

17.6.3 Characteristics of the Illustrated System

The illustrated concept in Fig. 17.6 assumes that the commercial module produces 2 megawatts of power at 35 gigahertz from a 40-meter phased array transmitter and antenna. The identical power system can be used for jamming application with the commercial phased array transmitter and antenna replaced by one tailored for the frequency bands to be used for jamming. For functioning as a power supply, only the power generating and conditioning elements would be used. The illustration shows a space-fed lens antenna as the power radiator, though other antenna types could be used as well.

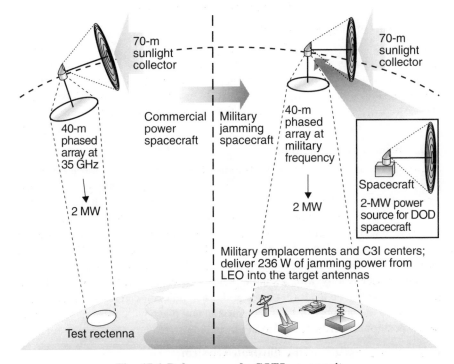

Fig. 17.6. Defense uses of a COTS power unit.

17.6.4 Performance and Weight

The jammer produces 200 terawatts of effective radiated power. From LEO it could deliver 256 watts of power (+24 decibel watt) into a receiver through the sidelobes of its antenna. Even if a large amount of antijamming processing gain were being used, there would be complete saturation of many receiver front ends. The capabilities from MEO and GEO are smaller due to the increased distances and the inverse square law, though still sufficient to overwhelm many electronic sets, some through their sidelobes. In this case raising the system altitude trades longer access time for less delivered jammer power.

Effectiveness calculations have not been performed as they depend on the characteristics of the system to be defeated. The power module alone would weigh 4000 kilograms. (If implemented using Buckytubes throughout, its total weight would be reduced in the far term to about 40 kilograms.)

17.6.5 Technologies and Time Frame

The principal technologies to be matured are adaptive membranes, MEMS FEEP thruster arrays, high energy thermal dissipation, and high power transmitters and antenna arrays at 35 gigahertz. These technologies probably could be demonstrated together by 2015 and the system deployed by 2020, making this a far term concept.

17.7 Wireless Energy Distribution Satellite

17.7.1 Concept Purpose and Utility

This concept presents a worldwide space power distribution network in which a few prime mover power sources distribute small to medium levels of power to a large number of utility customers. Its implementation would create a commercial power network. Competition would eventually result in several such global power nets distributing power to developing nations, and could be netted to form a a global power shared grid.

17.7.2 Principles of Operation

Power is beamed from a ground RF transmitter and reflected from an orbiting phased array reflector "reflectarray" and sent to ground rectennas. The phased array is phase controlled by command to dwell on a number of such rectenna sites in sequence as it moves in its orbit. The power transmitted is thus time shared, and the average level delivered at a rectenna is the transmitted power divided by the number of sites served. A practical scheme might scan the beam, dwelling on each receiver site exactly 1/120 or 1/100 of a second. Then each site can produce 60 or 50 hertz power readily with a simple polarity switch, so that the power can be impressed directly on the local power grid. Each reflector can relay power from a local source to a number of local rectennas in any place within line of sight of its orbital position, covering the globe in time.

17.7.3 Characteristics of the Illustrated System

The spacecraft consists of an adaptively flattened piezoelectric membrane on which are placed phase-shifting retrodirective elements. These are commanded from an onboard computer under the control of a ground master system to reflect an incoming beam toward one rectenna at a time. MEMS FEEP propulsion units are dispersed along the periphery of the film reflector and function for attitude control, translation control, and stationkeeping. The phased array elements as well as the MEMS FEEP thrusters are powered directly by absorbing a small fraction of the incoming beam. Using 35 gigahertz as the carrier frequency and a 100-meter-diameter reflectarray in a 6000–kilometer altitude orbit, the transmitter antenna as well as the rectennas can be as small as 1 kilometer in diameter. (See Fig. 17.7.)

17.7.4 Performance and Weight

The spacecraft can supply peaking power during a demonstration phase. Eight spacecraft in 6000-kilometer orbits would supply continuous baseband power to sites anywhere globally within a 50-degree latitude band. Assuming a 1000-megawatt power source, its entire power could be supplied to a single site, or 100 megawatts to each of 10 sites, or 10 megawatts to each of 100 sites within view. The spacecraft weighs approximately 3900 kilograms. (If implemented using Buckytubes throughout, its total weight would be reduced in the far term to about 40 kilograms.)

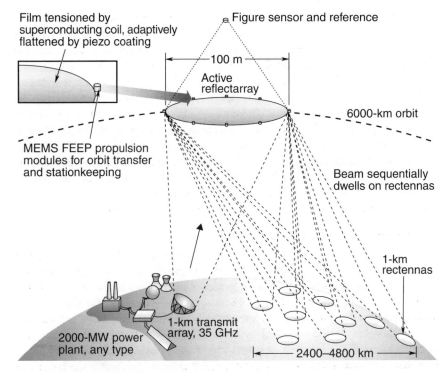

Fig. 17.7. Wireless energy distribution satellite. With one site per power plant, power to the site is 1000 MW; for 10 sites, power to each site is 100 MW; and for 100 sites, the power will be 10 MW to each site.

17.7.5 Technologies and Time Frame

The principal technologies required are those of adaptively shaped piezoelectric membranes; MEMS FEEP thrusters; phase-shifting, inexpensive 35-gigahertz modules; high power ground-based 35 gigahertz transmitter and antenna phased array elements; and matching rectennas. These technologies probably could be demonstrated in the 2015 time frame and deployed in the 2025 time frame, making this a far term concept.

17.8 LEO/MEO Solar Power Satellite

17.8.1 Concept Purpose and Utility

This concept presents means to provide electricity to Earth from space on a massive scale. Its implementation would provide wholesale quantities of clean and inexhaustible electrical power generated in space to Earth power grids worldwide, including developing nations, at advantageous prices.

17.8.2 Principles of Operation

This concept is an updated version of one of the principal outputs of the NASA Fresh Look Space Solar Power study, commissioned in 1996 by the author while at NASA. It revised the previous Solar Power Satellite "reference system," developed new architectures, and introduced new technologies in order to reduce weight, size, and cost, making the concept more competitive. The subject concept employs a modular construction with a large number of replicated power modules to reap the cost benefits of mass production. The power units use Fresnel membrane concentrators and spectrally split solar arrays, and are themselves arrayed on a gravity gradient tether. The transmitter is a space-fed lens with an adaptive membrane-supported phased array to generate and transmit the microwave power beam to the ground. Each spacecraft is in LEO or MEO orbit, and several are required for continuous coverage of the receiving sites. The users get power via a rectenna at each user site. The spacecraft is assembled and serviced robotically in space.

17.8.3 Characteristics of the Illustrated System

The system illustrated in Fig. 17.8 uses 5.8 gigahertz for power transmission. The spacecraft is 5 kilometers long, stabilized by the gravity gradient. It generates 400 megawatts of power, which it delivers to rectennas in either a dedicated or shared mode. The transmitting phased array on the spacecraft lens is 400–1200 meters in diameter depending on the orbital altitude chosen. The rectennas are 1–5 kilometers in diameter, also determined by orbit altitude. The tether replaces conventional structural trusses and supports the high power distribution bus. The transmitting antenna functions as a space-fed lens, with the feed at the lower end of the tether and the phased array antenna elements modulated for beam steering. Ion propulsion is used for attitude control and translation, and also enables self-ferry from LEO to its final orbit.

17.8.4 Performance and Weight

A few spacecraft could supply peaking power, and 18 spacecraft would provide continuous base load power to hundreds of rectenna sites, distributing their individual 400 megawatts of power. The system investment is an order of magnitude lower than that of previous concepts, as is the cost of the delivered power. The entire spacecraft weighs 900,000 kilograms in orbit. (If implemented using Buckytubes throughout, its total weight would be reduced in the far term to about 9000 kilograms.)

17.8.5 Technologies and Time Frame

The technologies required are those of large adaptive membranes shaped by electron beam, large ion engines, large Fresnel thin film concentrators, high power yet low cost phase-shifting antenna modules, high power transmitter elements, long fail safe tethers, low loss high power distribution systems, large capacity thermal dissipation systems, and others. These technologies probably could be demonstrated by 2015 individually and by 2020 collectively in space. A system could be deployed in the 2025–2030 time frame, making this a far term concept.

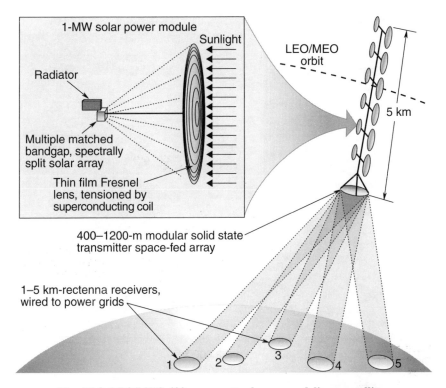

Fig. 17.8. LEO/MEO 400-megawatt solar power delivery satellite.

17.9 Provision of Energy to Lunar Bases

17.9.1 Concept Purpose and Utility

Solar power spacecraft in lunar orbit in this concept provide abundant electrical power to bases and outposts on the moon, during the lunar day as well as the lunar night.

17.9.2 Principles of Operation

A solar power converting spacecraft is stationed at the equigravity point between Earth and the moon. (See Fig. 17.9.) The spacecraft is placed into a stable halo type orbit around the Lagrangian equigravity point. The spacecraft is in sunlight almost continuously and can beam microwave or millimeter wave energy to rectennas on the moon even when they are in the darkness of the lunar night. The spacecraft needs little stationkeeping and will be placed into an inertial rotation of once per 28 days so as to be constantly moon pointing. The tether backbone for this spacecraft can be thousands of kilometers long yet very lightweight as the gravity gradient is extremely weak around this lunar point. Millimeter waves can be readily used for the transmission because there is no atmosphere, and their use will keep the size of the rectennas and the transmit array small. An alternative is to

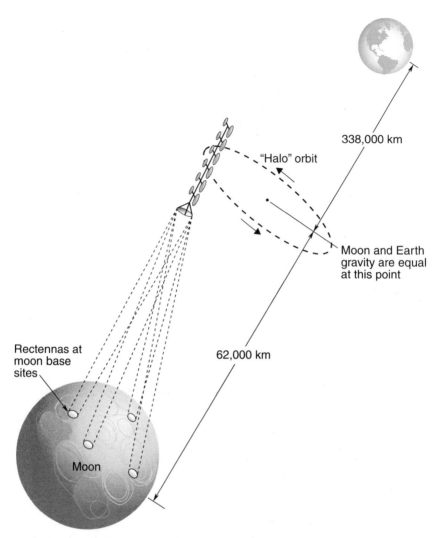

Fig. 17.9. Provision of energy to Lunar bases.

use a laser transmitter and a photovoltaic receive array, which are also feasible due to the absence of clouds or atmosphere in the propagation path.

17.9.3 Characteristics of the Illustrated System

The system envisioned is similar to that presented in the previous concept. The location of the equigravity point is approximately 62,000 kilometers above the moon's surface, so that the rectennas on the moon need to be 2 kilometers in diameter. Four hundred megawatts are delivered to the lunar outposts or bases, shared among however many sites exist. New sites can be added and the power redistributed with time by simple software modifications. A clear line of sight exists to Earth at all times for communications and control.

17.9.4 Performance and Weight

The system is very similar to that described in the previous concept for LEO/MEO application. A developmental power unit could be transferred to this location after completing its test program in LEO. The weight of the spacecraft is identical to that of the previous concept.

17.9.5 Technologies and Time Frame

The technologies required are those of large adaptive membranes shaped by electron beam; large ion engines; large Fresnel thin film concentrators; high power, low cost phase-shifting antenna modules; high power transmitter elements; long fail safe tethers; low loss, high power distribution systems; large capacity thermal dissipation systems; and others. These technologies probably could be demonstrated by 2015 individually and by 2020 collectively in space. A system could be deployed in the 2025–30 time frame, making this a far term concept.

17.10 GEO Space Solar Power Spacecraft

17.10.1 Concept Purpose and Utility

This concept presents means to provide truly massive quantities of clean and inexhaustible electrical power generated in space to power grids worldwide at competitive or even advantageous prices compared with all alternatives. Its adoption would allow the rapid industrialization of much of the underdeveloped world.

17.10.2 Principles of Operation

This concept is an updated version of one of the principal outputs of the NASA Fresh Look Space Solar Power study. A number of concepts are presented, all operating from GEO. They range from solar photovoltaic power generation using sun-oriented thin film photovoltaic arrays and Earth-oriented space-fed lens microwave transmitting arrays, to sun-oriented thin film mirrors illuminating despun photovoltaic arrays attached to space-fed lens microwave transmitting arrays. In each case the transmitter antenna lens consists of an adaptive membrane supporting many phase shifter modules to generate, transmit, and steer the beam to the ground. The beam can be dedicated to one rectenna site or hop among several rectennas, distributing its power among them. The spacecraft is in GEO and so has continuous coverage of much of its visible hemisphere on the Earth. The spacecraft are macrostructures that are assembled and maintained robotically, and grow in capability incrementally.

17.10.3 Characteristics of the Illustrated Systems

Two systems are shown in Fig. 17.10, both using 5.8 gigahertz for power transmission. Both generate power starting at 400 megawatts and growing to 10 gigawatts

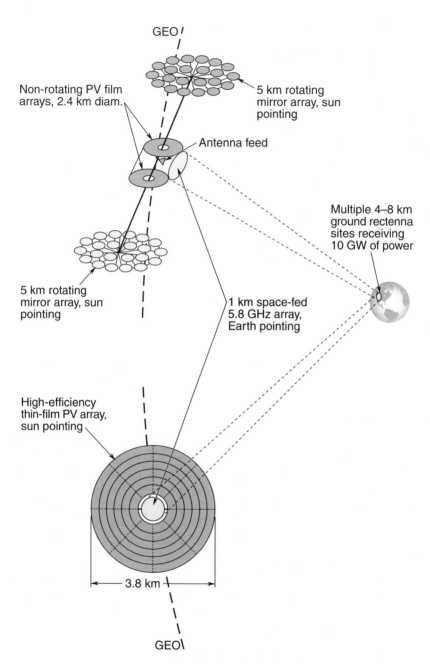

Fig. 17.10. Two alternative GEO space solar power spacecraft.

with the addition of modules. The transmitting lens is 1 kilometer in diameter, and the rectennas are 4 to 8 kilometers in diameter. The direct photovoltaic system uses large rotating contact joints to decouple the antenna from the rotating solar array, while the optical system avoids such rotating contacts by using only rotating mirrors. The system for attitude control and translation is ion propulsion. Both systems also ferry themselves from LEO to their final orbit in GEO. These are massive construction projects in space on a scale never before attempted. Each of these concepts has had preliminary designs funded by NASA, and both have been found feasible, though challenging.

17.10.4 Performance and weight

Each spacecraft could supply continuous base-load power to one receiving site at 10 gigawatts or to 10 sites at 1 gigawatt each. The system investment is an order of magnitude lower than that of previous GEO concepts. The cost of the delivered power is also an order of magnitude lower and less than that of the LEO/MEO power concept, and is competitive if not advantageous compared with conventional power sources, particularly since its delivered energy price need not be penalized for environmental pollution. Both spacecraft types are on the order of 4–5 kilometers across and weigh on the order of 25 million kilograms in orbit. (If implemented using Buckytubes throughout, their total weight would be reduced in the far term to about 250,000 kilograms each.)

17.10.5 Technologies and Time Frame

The technologies required are those of large adaptive membranes shaped by electron beam; large ion engines; high power, low cost phase-shifting antenna modules; high power transmitter elements; long fail safe tethers; low loss, high power distribution systems; large capacity thermal dissipation systems; and large lightweight truss structure elements in some concepts. These technologies probably could be demonstrated by 2015 individually and by 2020 collectively in space. A system could be deployed beginning in the 2025–2030 time frame, making this a far term concept.

17.11 RF Powered Electromagnetic LEO Launch System

17.11.1 Concept Purpose and Utility

This concept presents means to implement a launch vehicle that requires almost no onboard propellants. Its implementation promises to reduce costs of space launch by three orders of magnitude from today's levels, even without using Buckytubes. It is one of a class of vehicles that, if implemented, would create an industrial revolution in space due to the low costs of access to space. This concept was described in the transportation section (Section 16.16) but is repeated here to emphasize that its heart is a power beaming concept. The concept is illustrated in Fig. 17.11.

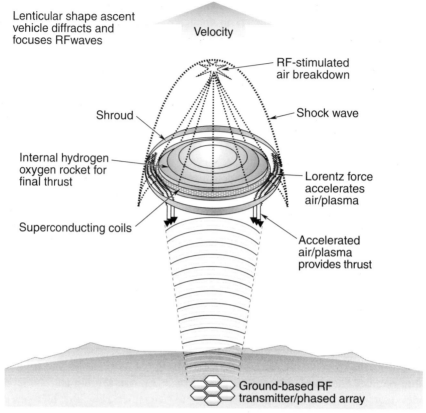

Lenticular shape ascent vehicle diffracts and focuses RFwaves

Velocity

RF-stimulated air breakdown

Shroud

Shock wave

Internal hydrogen oxygen rocket for final thrust

Lorentz force accelerates air/plasma

Superconducting coils

Accelerated air/plasma provides thrust

Ground-based RF transmitter/phased array

Fig. 17.11. RF-powered electromagnetic LEO launch system.

17.11.2 Principles of Operation

Millimeter wave or microwave energy is generated on the ground and beamed to the launch vehicle. There it diffracts around the shaped vehicle and is focused at a point ahead of the vehicle, causing atmospheric breakdown due to the intense energy density at that point. The resulting ionized air flows downward around the vehicle and is accelerated by superconducting coils at its base using magnetohydrodynamic forces. This accelerated air produces downward thrust akin to a jet engine. Since the reaction mass is air, there being no onboard propellants, and the energy source is on the ground, the vehicle specific impulse is infinite, and launch energy costs stem principally from the costs of electricity to power the beaming system. Nonetheless a relatively small internal source of hydrogen is used as the propellant in a rocket mode for the final ascent to orbit when out of the atmosphere.

17.11.3 Characteristics of the Illustrated System

The accelerated air is ejected through nozzles at the periphery of the vehicle's base. Multiple superconducting coils used in groups around the periphery of a lenticular-shaped vehicle allow hovering and translation sideways much as a helicopter. The vehicle's coils act together to propel the vehicle toward orbit. The ground system comprises a power supply, high power microwave or millimeter wave beam, and a phased array antenna to form and propagate a narrow beam toward the spacecraft. Automated tracking and safety systems would ensure controlled energy flow to the vehicle. The system is completely reusable and consumes no propellants except for the hydrogen and oxygen used for final ascent to orbit, which use the same circular slot as an exit nozzle as is used by the ionized air. A laser-powered system in which a ground laser caused air breakdown and propelled a passive aerodynamic shape upward has already been tested at small scales and low velocities, demonstrating some of the required principles.

17.11.4 Performance and Weight

This system has been proposed and analyzed by Leik Myrabo of Rensselaer Polytechnic Institute and supported by NASA. It promises eventually to result in launch costs of $20 per kilogram to orbit. (If implemented using Buckytubes throughout, its dry weight would be reduced in the far term by a factor of 100, and its operating costs would be reduced in the far term by a factor of 10 to about $2 per kilogram.)

17.11.5 Technologies and Time Frame

A number of technologies require development, including the high energy and power beaming millimeter or microwave system, the superconducting coils for the vehicle and their controls, and the vehicle design to satisfy aerodynamic, magnetohydrodynamic, and chemical rocket requirements. These technologies probably could be demonstrated acting together in the 2020 time and a vehicle deployed in the 2025–2030 time frame, making this a far-term concept.

17.12 GEO Mirror, Ground Laser Force Projection

17.12.1 Concept Purpose and Utility

This concept describes a ground laser used in conjunction with a space mirror as both a surveillance system and a directed energy weapon against ground, air, and space targets. Its implementation would provide missile defense, space object tracking, air defense surveillance, and space defense. It would also have a self-defense capability. The system is illustrated in Fig. 17.12.

17.12.2 Principles of Operation

A reflecting optical wavelength phased array (reflectarray) is positioned in GEO. The array is supported by a piezoelectric membrane whose shape is adjusted by an electron beam in response to an optical figure sensor. A second stage of correction modulates the optical phase shifters to remove the effects of residual surface errors and attain an optical quality reflection of the ground beam. A

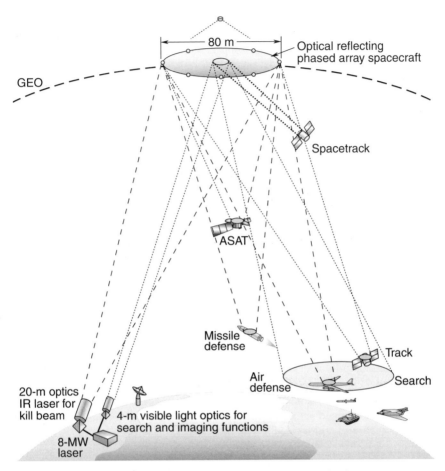

Fig. 17.12. GEO mirror, ground laser force projection.

two-dimensional angle of reflection can be modulated in real time by adjusting the phases of the phase-shifting elements.

The mirror can be made flat throughout or can have a superimposed smaller portion that can be curved or whose planar tilt can be separately commanded and set. This latter portion is viewed by a CONUS-based optical sensor and establishes a surveillance capability with its sensing angle modulated electronically. The entire surface is illuminated by a ground-based high energy laser and can direct its reflected beam onto any desired target by modulation of the phases of its elements, by changing the attitude of the entire reflector, or both. The high energy laser beam is steered by ground command in response to cueing and to signals obtained from the surveillance capability and other inputs.

The system covers nearly a hemisphere on the Earth as well as in space. It can defend itself against spacecraft interceptor threats. It can perform the missile defense functions from a CONUS base. Three or four space systems would be needed for global capability.

17.12.3 Characteristics of the Illustrated System

The optical phased reflectarray in GEO must have a diameter of 80 meters in order to produce a 1-meter laser spot on the ground or air. The modulatable portion of the reflector is 2 meters in diameter in order to have a 1-meter resolution element on the ground, and its curvature, location, and tilt with respect to the main reflector can also be adjusted by command. The ground system consists of a visible light surveillance optical sensor with a 4-meter diameter aperture and an 8-megawatt infrared laser with 20-meter diameter aperture for the kill function.

17.12.4 Performance and Weight

The system is only notionally designed as some technologies require considerable advances before even a preliminary system design can be completed. Based on scaling of other adaptive optical reflector concepts, the reflectarray will weigh about 6000 kilograms. (If implemented using Buckytubes throughout, its total weight would be reduced in the far term, in principle, to about 60 kilograms.)

17.12.5 Technologies and Time Frame

The principal technologies required are those of piezoelectric adaptive electron beam-shaped membranes and massively replicated optical phased array elements, both operable at high incident energy densities and capable of nanometer coherent accuracies; MEMS FEEP thrusters; and ground-based, high energy lasers. These probably could be demonstrated in space working together by 2020–2025 and the system deployed by 2025–2030+, making this a far term concept.

17.13 Ultrafast Interplanetary Transport

17.13.1 Concept Purpose and Utility

A novel propulsive transfer vehicle concept is described that shortens the travel time to the major planets in the solar system from many years to a few weeks. Its implementation would make major reductions in life requirements for components, radiation damage to humans and equipment, and total mission costs.

17.13.2 Principles of Operation

An inertial confinement fusion engine, shown in Fig. 17.13, is used, which implodes a series of small deuterium-tritium or deuterium-helium-3 pellets. The implosions result in bursts of fusion-released energy, which heat and accelerate hydrogen fuel and propel the spacecraft. The fusion principle does not need to attain energy breakeven and has already been demonstrated at Lawrence Livermore National Laboratory in Livermore, California.

The biggest problem with this type of fusion engine is the many ultrashort pulse lasers needed to illuminate the pellet from all directions and cause it to implode, as

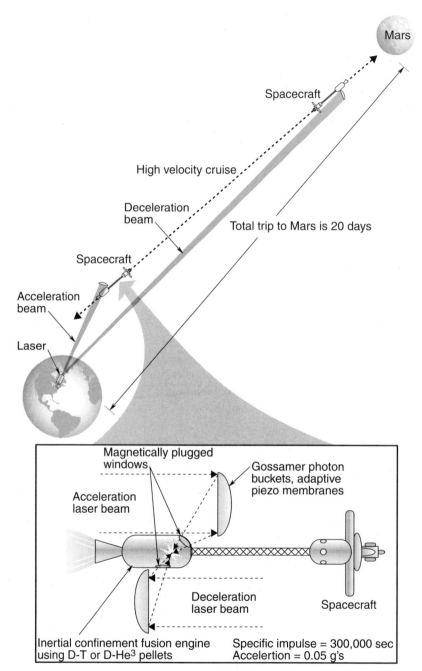

Fig. 17.13. Ultrafast interplanetary transport: inertial confinement fusion with beamed implosion energy.

these lasers are extremely large, heavy, and expensive. This concept leaves the lasers on the ground and beams their energy to the space vehicle only when needed. The laser energy is collected on board the spacecraft by a large but imprecise photon bucket, split into the requisite many beams, and focused on the pellet to be imploded through a magnetically plugged window. Many tiny imploding pellets provide rapid propulsion pulses, which average to the desired thrust.

After a period of implosions for acceleration, the spacecraft coasts at high velocity for an extended period. A second such collector is used for the deceleration maneuver, or alternatively the vehicle can be turned around to fire forward and just the collector reversed. In either case the deceleration collector must be larger than the acceleration collector because the spacecraft is further from Earth at use. This concept will attain a specific impulse of 300,000 seconds, thus requiring roughly 1000 times less propellant mass than a chemical propulsion system would require.

17.13.3 Characteristics of the Illustrated System

For a mission to Mars the laser transmitter on the ground has a 10-meter diameter and a laser power of 10 megawatts. The photon bucket on the spacecraft needs to be 500 meters in diameter for the deceleration propulsion. Its optical quality can be poor, however, as the beam needs only to be focused to a diameter on the order of a millimeter rather than a micron, and therefore an adaptively corrected piezoelectric membrane without a second stage of correction will be sufficient for this application.

Figure 17.13 shows two photon buckets for simplicity. The acceleration of the vehicle will be a constant 0.05 g during engine operation. The coast velocity will be 50 kilometers per second. The propellant mass needs to be only 6 percent of the total vehicle mass to provide both acceleration and deceleration, compared to greater than 90 percent for chemical propulsion. The power, collector size, and coast velocities all increase to reach more distant planets, as shown in Table 17.2.

Table 17.2. Energy Cost per Trip to Three Planets

Specifications	Mars	Jupiter	Saturn
Trip time (days)	20	60	90
Coast velocity (km/sec)	50	150	200
Laser diameter (m)	10	10	10
Photon bucket diameter (m)	500	1000	2000
Propellant mass fraction	0.03	0.1	0.13
Laser power MW	10	50	100
Energy cost per trip ($million)	20	100	200

17.13.4 Performance and Weight

Total travelling time to reach and rendezvous with Mars is 20 days; Jupiter, 60 days; and Saturn, 90 days. The energy cost is minuscule compared with equivalent costs of any planetary mission, and far smaller than those for any other form of propulsion.

17.13.5 Technologies and Time Frame

The technologies needed are those for inertial confinement fusion, magnetic engine nozzles, ultrashort high power and energy lasers, adaptive electron beam-shaped piezoelectric materials for the photon bucket collecting apertures, and magnetically plugged windows—all operable in a high peak power energy environment subject to radiation. These probably could be demonstrated by 2020–25 and deployed by 2025–30, making this a far term system.

17.14 Geostationary LEO/MEO Power Relay Using Buckytubes

17.14.1 Concept Purpose and Utility

A spacecraft is suspended at low or medium altitudes over a point on Earth's surface and is therefore geostationary though it is not in the conventional GEO. The spacecraft can perform all the usual functions of GEO spacecraft, but with antennas or optics 1/10 to 1/100 the size or powers 1/100 to 1/10,000 as large. One such spacecraft can provide base load power to all receiving sites within its coverage area.

17.14.2 Principles of Operation

The development of unidirectionally aligned Buckytube strings will probably precede other structural shapes. These strings will be assembled into long space tethers of remarkably low weight for a given strength. A tether extending from near the Earth's surface to GEO will weigh no more that the payload that it is to support. This is in stark contrast to the tether mass if current materials were to be used, which would weigh an impractical 100,000 times more than the payload to be suspended.

Thus a reflectarray can be suspended at, say, 400-kilometer altitude and reflect surface generated power to receivers within a 2000-kilometer radius area using transmit antennas and receiving rectennas 1/20 the size of those that would be required had the reflectarray been in GEO. The "spacecraft" consists of the payload, the tether, and a countermass above GEO that can consist of dead GEO satellites. The "spacecraft" is in static balance, with the center of mass of the entire spacecraft in GEO. The system operates in exactly the manner of the LEO/MEO reflectarray system discussed in the concept of Section 17.7.

17.14.3 Characteristics of the Illustrated System

An example, shown in Fig. 17.14, illustrates a power relay system that is suspended at various geostationary altitudes. A 50-meter diameter reflectarray is used with a 200-meter transmit antenna diameter and rectennas only 300 meters in diameter. These are small enough that they are much more affordable for even small power installations and receive sites, and will raise fewer objections to installation. Each

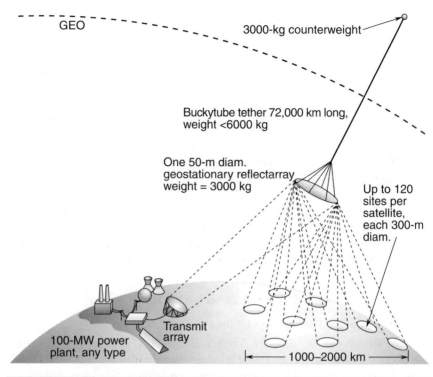

Fig. 17.14. Geostationary LEO/MEO power relay using Buckytubes.

Orbit Altitude	Transmit Size	Receive Size	Coverage	Number of Sites	Power per Site	Typical Location and Coverage
1000 km	320 m	300 m	2400 km	up to 120	1 MW min	Indonesia, Borneo, Malaysia, 70% of Philippines
3200 km	1 km	1 km	4800 km	up to 120	1 MW min	Mexico, Central American, 1/2 South America

prime power plant could distribute energy to up to 120 receive sites, providing 60-hertz power to all. Because the spacecraft does not move with respect to the surface although at low altitude, one such spacecraft can now provide continuous base load power to a whole region, much as do surface wire transmission systems. Many coverage areas over the globe can be implemented using other such spacecraft. Placing such a spacecraft into inclined orbit, yet with a 24-hour period provides a larger ground coverage, as the subsatellite point describes a figure 8, at the possible expense of supplying only noncontinuous power to some rectennas, or requiring multiple such systems for continuous power delivery to all sites.

17.14.4 Performance and Weight

Power can be transmitted from one or a number of prime sources to at least 120 receiver rectenna sites within a 2000-kilometer radius area. Greater coverage together with larger reflectarray, transmit, and receive antenna sizes results if the suspension altitude is raised, and the tether weight reduces rapidly with such altitude increases. The weight of a LEO/MEO geostationary power relay with the tether constructed entirely from Buckytubes is 12,000 kilograms, 3000 kilograms of which are dead satellites.

17.14.5 Technologies and Time Frame

The principal technologies required are those of long-life, very long Buckytube tethers, adaptively shaped piezoelectric membranes, MEMS FEEP thrusters, inexpensive phase-shifting reflectarray modules, high power ground-based 35-gigahertz transmitter and antenna phased array elements, and large rectennas. These technologies probably could be demonstrated in the 2020–25 time frame and deployed in the 2025–30 time frame, making this a far term concept.

17.15 Aircraft Powering Via Laser Beams

17.15.1 Concept Purpose and Utility

Large aircraft can be remotely powered so that they can stay aloft indefinitely without requiring fuel, have global range, or both. Implementation would result in much lower aircraft operating costs, almost complete elimination of their atmospheric pollution, and the ability to undertake entirely new missions.

17.15.2 Principles of Operation

Large yet lightweight optical reflectors in space reflect ground-generated, high energy laser beams to aircraft in flight. At the aircraft, the energy is collected by optical apertures on or in the wings and focused into the jet engines. Inside the engines the high energy density heats and breaks down the air, which then flows out through the turbines at high speed in the same manner as a combusted fuel air mixture, furnishing forward thrust. A small amount of jet fuel would be carried to allow for takeoffs and landings in the presence of clouds, as well as climbout and descent through cloud cover, but at high altitude the aircraft would be able to cruise indefinitely without fuel. The concept is illustrated in Fig. 17.15.

A large transport aircraft requires about 50 megawatts of power in the cruise mode. The mirror spacecraft could power a number of aircraft by beam hopping among them in sequence, the pulsed energy thus received in each aircraft having a high pulse rate such that smooth averaged thrust results. Alternatively each spacecraft could consist of a collection of reflectors so that the aircraft has a dedicated mirror for its flight in the coverage region of each spacecraft. In either case the mirrors would be steerable. The beam's very existence would be made dependent on the reception of a pilot beam originating on the aircraft to ensure that the beam

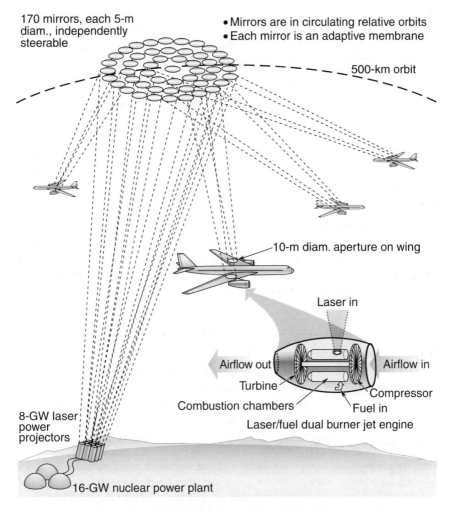

170 mirrors, each 5-m diam., independently steerable

• Mirrors are in circulating relative orbits
• Each mirror is an adaptive membrane

500-km orbit

10-m diam. aperture on wing

Laser in

Airflow out   Airflow in

Turbine

Combustion chambers   Compressor
                              Fuel in

8-GW laser power projectors

Laser/fuel dual burner jet engine

16-GW nuclear power plant

Fig. 17.15. Remote powering of aircraft.

ceased to exist if it veered from the airplane's collector for any reason. This could be ensured by placing the spacecraft mirror and the aircraft collector inside the lasing loop, with the ground system being a laser amplifier. Since the mirrors would orbit at low altitudes to minimize aperture sizes, a number of spacecraft clusters would be needed for global coverage.

17.15.3 Characteristics of the Illustrated System

The spacecraft consists of 170 mirrors, each 5 meters in diameter. These are placed in circulating relative orbits about a central mirror, and require only second order primary stationkeeping propulsion. An 8-gigawatt laser cluster on the ground focuses laser energy on the spacecraft mirror cluster, operating from a nuclear prime source. The spacecraft cluster in 500-kilometer orbit places a 10-meter diameter laser spot on the same size collector aperture on the aircraft wing and powers 170 aircraft simultaneously, one per cluster mirror.

17.15.4 Performance and Weight

The aircraft can stay aloft as long as there is a clear line-of-sight path, its altitude is above about 3000 meters, and the ground laser functions normally. The aircraft can switch to onboard fuel at any time for seamless power, which it can also use for takeoffs and landings and operations under cloud cover. All the world's large aircraft could be powered by a number of such constellations of reflector spacecraft, creating a global aviation power grid.

17.15.5 Technologies and Time Frame

The technologies that need to be matured include adaptive electron beam-shaped piezoelectric mirrors operable at high incident energies, FEEP thrusters for attitude control and stationkeeping, extremely high power ground lasers, compact optical collectors for the aircraft, and dual mode jet engines. These probably could be demonstrated by 2025 and deployed by 2030+, making this a far term concept.

17.16 Interstellar Probe Experiment

17.16.1 Concept Purpose and Utility

This concept presents means to undertake the first experimental star probe. Its implementation would prove the technology to be eventually used to send a scientific probe to the vicinity of a nearby star. It is the most that can be accomplished in this area in the time frame of this book.

17.16.2 Principles of Operation

The probe would use a large gossamer sail that would be accelerated by the reaction of a high power laser or microwave beam from Earth. A high power source of energy with an extremely large transmitting aperture would be required to illuminate the sail by the beam from Earth at stellar distances with little spillover. An extremely large sail would be required to intercept most of this beam, particularly as the sail distance increased. The sail would accelerate as long as it was in the beam, achieve a fraction of the speed of light, and then coast to fly by a star system while sending information back to Earth. An actual scientific probe would have to attain 10 percent of the speed of light to have a transit time of "only" 50 years, the maximum reasonable given the lifetimes of scientists. Such probe would require a 1000-terawatt laser on Earth, and a sail with a diameter of 1000 kilometers. It would attain a specific impulse of 10 million. Because of these daunting numbers, this concept therefore concentrates on a technology demonstrator probe with less demanding requirements. The concept is illustrated in Fig. 17.16.

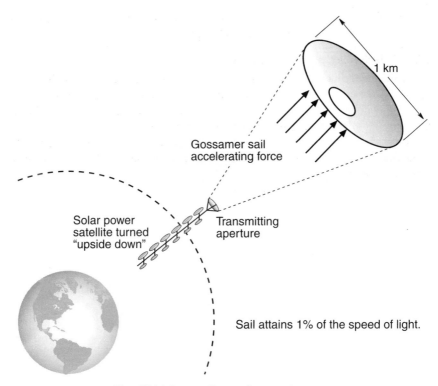

Fig. 17.16. Interstellar probe experiment.

17.16.3 Characteristics of the Illustrated System

The system is an experimental probe that achieves only 1 percent of the speed of light. It would take 1000 years for the journey to a nearby star at those velocities and hence is only a technology demonstrator, which is not expected to be operated for more than a few dozen years. It also uses microwaves instead of a laser, capitalizing on a solar power spacecraft in Earth orbit turned "upside down" to transmit power to the sail. The sail consists of a gossamer material incorporating electronic chips in its very fabric to implement an attitude control function and also function as a distributed transmitter for sending telemetry to Earth. The sail consists of a loose mat of carbon fibers weighing only 0.001 kilograms per square meter of area.

17.16.4 Performance and Weight

The sail is 1 kilometer in diameter. It uses the 10 gigawatts produced and projected by the space solar power spacecraft in Earth orbit described in an earlier concept, and demonstrates the achieving of scaled acceleration, specific impulse, and control techniques. It weighs less than 1000 kilograms when constructed as a mat of conventional carbon fibers. When made entirely of single wall Buckytubes, the weight of the sail will approach 100 kilograms.

17.16.5 Technologies and Time Frame

The principal technologies to be matured are adaptive shape-controlled ultrawispy materials, control techniques for such floppy membranes, integrated distributed very lightweight electronics, and MEMS FEEP thruster clusters in addition to those for the solar power satellite. These probably could be demonstrated by 2025 as could the space solar power satellite for producing the power beam. The deployment of this experimental probe could then take place in 2030+ and thus is a far term concept.

Part IV

Assessments and Conclusions

18 Observations

The applications presented in Part III have such broad scope and represent such different scales of activities in different time periods that some unifying observations and broad perspectives are imperative to prevent the forest from being lost for the trees. The principal observations, given in Table 18.1, are discussed in this chapter.

18.1 Adaptive Gossamer Materials

One of the most powerful techniques advanced in this report is the use of adaptive gossamer membranes, uninflated, unsupported, and adaptive along their entire surfaces, to form large, precisely shaped reflectors for radio frequency (RF) and optical sensors. These membranes would be formed of piezoelectric bimorphs or the like and actuated by external electron beams in response to external figure sensors observing their actual figure. This closed loop action forms the desired figure only when the device is finally in space and maintains it in response to software in the presence of external or internal disturbances. Thus, rather than depending on extreme precision of manufacture and rigidity for handling and deployment to attain a given reflector figure, this new technique forms and maintains the desired figure, both in the large and fine scale, using information rather than structural rigidity. The result is inevitably a much lighter and cheaper system capable of much greater apertures with greater precision than can be practically attained using mechanical precision.

Although surface accuracies suitable for RF use are attainable directly by this technique, attaining surface accuracies suitable for optical applications is orders of

Table 18.1. Some Observations on High Leverage Technologies

- Adaptive gossamer materials will enable very large yet extremely lightweight filled aperture antennas and optics.
- Coherently cooperating swarms will enable even larger optical and RF arrays.
- Buckytube materials will reduce the weight of everything by a factor of 100.
- Launch to orbit and propellantless orbit transfer will be almost cost free.
- Extremely small as well as extremely large space systems will exist, and they will both be lightweight and inexpensive.
- Spacecraft will look distinctly different from how they look today, or would look simply if extrapolated using current technologies.
- Spacecraft will be routinely serviced and upgraded, in orbit and on the ground.
- Spacecraft will also be proliferated, inexpensive, and economically expendable.
- Brassboards and many spares will be routinely orbited.
- Megawatt power levels will be inexpensive and widely available in space.
- Powerful information transmission and processing will meet all demands.

magnitude more difficult. A second stage of correction is used, therefore, in which the remaining surface irregularities are corrected in a liquid crystal spatial light modulator, which then yields an optically coherent output with good imaging qualities.

These gossamer membranes shaped by closed loop electron beams are free standing in space, untensioned and uninflated, controlled in attitude and translation by MEMS FEEP thruster arrays around their periphery, and not attached by structural trusses to other instrument or spacecraft components. All major components of the spacecraft are thus stationkept or formation flown with respect to each other, eliminating the large and heavy trusses required in conventional sensors to hold the primary aperture precisely with respect to the secondary aperture or focal arrays and detectors.

With these techniques the size and weight of a large aperture sensor spacecraft will be three to four orders of magnitude lower than with current techniques, and one to two orders of magnitude lower than with any other advanced technique identified to date including inflatable membranes. In addition, the sensor aperture need not be limited by the size of a launch vehicle's payload shroud because the membrane can be tightly folded for launch, as the closed loop figure control and correction technique will remove any creases remaining after deployment.

This ability to develop and deploy extremely large yet extremely lightweight and inexpensive filled aperture RF and optical sensors will revolutionize scientific, commercial, and defense applications.

This is the result of one embodiment of the powerful principle discussed in Chapter 2 regarding replacing structures with information.

18.2   Coherently Cooperating Swarms

Another powerful technique for replacing structures with information is to place many formation flown small satellites into a loose "swarm" and cause them to cooperate coherently. This is very different from the so-called distributed small satellite LEO constellations currently pursued in which individual small spacecraft perform essentially the same functions as larger satellites but at lower satellite weight and lower cost. These smaller spacecraft can be proliferated to provide greater geographical coverage for the same cost. The lower weight also saves launch costs, so the total system costs less for the same function performed with larger spacecraft.

In contrast, the swarms, as described in Part II of this report, cooperate coherently and form a real distributed system in which the whole is more than the sum of the parts. A generic description would be a constellation of small spacecraft each performing its separate function, but these functions combine to create at a central location a much larger virtual spacecraft, or sensor aperture, that exists solely because of the cooperation of the spacecraft. Examples of implementation of such systems in this report cause each of a large number of small satellites in a swarm or other constellation to radiate or receive signals and combine them in phase, or coherently, regardless of their actual location. This creates coherent RF or optical apertures that are essentially unlimited in size.

In such use, each satellite's position is only crudely stationkept, and the satellites adjust the time delay or phase delay of the signals they repeat to compensate for their positional errors, causing their repeated signals to add coherently at a collection point. This technique can be applied easily to RF transmitters and receivers, and with more demanding accuracy to optical receivers and transmitters. The result in either case is a large "swarm" or loose constellation of satellites that act as one large antenna or optical array, even though they are separate and their positions are neither constant nor lie along a parabola or plane in space. The individual satellites can be as simple as one-element flying chips or as complex as today's self-contained sensing spacecraft of various sizes.

The advantages of such truly coherently cooperating distributed systems is that they can form sparse RF antennas and optical sensors with diameters so large that they would be impossible to implement with filled apertures even if formed with adaptive membranes; and have orders of magnitude smaller weight. The relative locations of individual spacecraft in the swarm can be controlled by MEMS FEEP propulsion, tethers, or by cleverly conceived orbits in which the elements of the array appear to orbit a common center within it, thus eliminating the need to use propulsion at least for first order stationkeeping.

These array functions can be made coherent over very great distances. RF antennas with sizes of hundreds of kilometers and optical telescopes of hundreds of meters diameter can be formed. These systems can indeed enable new capabilities not possible with unitary spacecraft either acting alone, as a proliferated but noncoherent constellation, or as relays for each other. Formed of grown rather than assembled picosatellites or femtosatellites, such swarms will contain so many spacecraft that the economics of true mass production will come into play in space for the first time, greatly reducing the cost of producing the system. In addition, these systems feature the advantages of truly distributed satellite systems, including fault tolerance, robustness, survivability, reconfigurability by software, and the ability to be incrementally emplaced and upgraded as budgets are available.

It is thus important to recognize that today's so-called "smallsat" revolution has barely begun, and that today's "distributed" space systems, which are mostly proliferated smaller satellites that act alone in their principal functions, must evolve into fully coherently cooperating swarms of spacecraft whose individual sizes devolve toward picosats weighing about 1 kilogram or femtosats weighing 100 grams or less.

18.3 Buckytube Materials

The advent of single wall carbon nanotubes (Buckytubes) consisting purely of carbon atoms connected only by atomic bonds and requiring no matrix materials to hold their alignments will revolutionize the manufacture of all structures, components, tanks, engines, plumbing, wires, packages, antennas, solar arrays, electronic boards, and other components. Essentially all components and subsystems from which spacecraft and launch vehicles are constructed and assembled will be able to be made from these Buckytube materials. The resulting weight reduction will be nothing short of revolutionary.

Buckytubes exhibit a strength-to-weight ratio 600 times greater than that of high strength steel or aluminum alloys. New technologies being pursued today will enable the growth of Buckytubes into structures able to withstand forces in many directions, and be grown directly into any number of net shapes of spacecraft components. These shapes will consist almost exclusively of atomic bonds and will avoid the use of a matrix such as used today for making carbon composites, though Buckytube composite materials will make their appearance earlier and indeed may be preferable for some structures such as compression members. Even if we reduce the eventual strength-to-weight ratios attained by net shapes grown entirely from Buckytubes to "only" 100 times greater than that of high strength steel or aluminum alloys, allowing for a factor of six strength loss due to sidewall bonds in the atomic structures and practical limitations due to minimum gauge and bending stress-limited structures, Buckytubes will change not only space but all Earth-bound structures.

Spacecraft that now weigh 2500–5000 kilograms will weigh 25–50 kilograms, as essentially all their components can be made from Buckytubes. Launch vehicles constructed out of Buckytubes will have such a low structural mass fraction, or conversely high fuel fraction, that they will weigh 10 times less on the pad fully fueled for the same payload performance, or if not made smaller will be able to place 10 times their current payload to orbit. Single stage fully reusable vehicles will become not only feasible but practical, highly efficient, and the preferred launch vehicle configuration. Even with no improvements in propulsive technique in chemical rockets, launch costs per kilogram of payload will be reduced by one order of magnitude from today's levels.

The net weight reduction resulting from adoption of Buckytubes alone is thus a factor of up to 1000: a factor of up to 100 from reduction of spacecraft weights and a factor of up to 10 from the increased capability of launch vehicles for the same size. This portends a real revolution.

While these are the expected ultimate results of the introduction of Buckytubes, history teaches that the ultimate is usually longer in coming than anticipated and more likely preceded by a number of incremental steps yielding incremental intermediate benefits. Thus composite materials will likely be improved gradually by the introduction first of unoriented Buckytubes bound by epoxy or other matrix materials, and later by oriented Buckytubes still bound by a matrix. While these composites may initially have only a factor of two to four increase in strength, stiffness, and reduction of weight from today's composites, they will still have tremendous impact as such factors are truly major steps in structural materials. Thus while the ultimate promise still comes from the increase of a factor of 100–600, it will be preceded by a number of smaller incremental strength/weight increase steps, each a minor revolution its own right.

In a sense the Buckytube revolution will be easier to accept as it occurs in this incremental fashion. Consider that a Hubble Space Telescope that weighs 12,000 kilograms today would weigh an estimated 25–50 kilograms in orbit if constructed of adaptive membranes and stationkept elements, but still using conventional materials. (This is an estimated downward scaling of a 25-meter-diameter optical imaging spacecraft design.) It will be hard enough for current spacecraft designers to accept the feasibility of such low weights, let alone the further likely major weight reductions if Buckytube materials were also used throughout. In that case in principle, if the effects of Buckytubes and other new technologies were truly additive, and minimum gauge and other manufacturing limitations were not limiting, the total in-orbit weight of a spacecraft the size of the Hubble Telescope would be 0.3 kilograms, a 25-meter-diameter imaging spacecraft would weigh 3 kilograms, and a filled aperture 200-meter-diameter optical imager would weigh less than 100 kilograms. Thus, while such extremely low weights could, in principle, be attained in the longer term future, even if they could be shown to be practically attained they would hardly be accepted today even as a future projection. This skepticism will only be overcome by incremental demonstrations and deployment of intermediate capabilities.

The high leverage techniques espoused in this book together with the wholesale use of net-shape grown Buckytubes as well as Buckytube composites—both of which are sure to happen although their schedules are not easy to predict—will bring about a complete revolution in space systems. These systems will be so far out of the experience and imagination of current planners, designers, and builders of spacecraft, that even their serious contemplation is sure to be derided and resisted. Incremental progress will make them easier to accept, which will clearly happen though the path will not be easy.

18.4   Launch to Orbit and Propellantless Orbit Transfer

Advances in launch vehicles brought about by the new technologies discussed in Chapter 2 will reduce launch costs by a factor of 1000 to the order of $20 per kilogram even without the introduction of Buckytubes. Initially mostly conventional materials will be used to build fully reusable vehicles, with automated checkout and operations contributing to their expected operating costs of $200 per kilogram. Eventually beamed energy from the ground, powered from the electrical power grid, will mostly replace onboard propellants, and many vehicles will be flown very frequently to satisfy the enormous market created by the applications envisioned in this report as well as the introduction of space tourism—a space natural that by itself will dwarf current space markets.

The net result of the introduction of these technologies and the resulting high flight rates will be that launch costs to orbit will fall by a factor of 1000 to $20 per payload kilogram. While this is dramatic enough, the concurrent introduction of Buckytubes for all elements of the launch vehicle structure, engines, pumps, electronics, actuators, and other components will reduce the vehicle's "dry" weight conservatively by another factor of up to 100. This translates directly into an increase in the payload orbited by a single stage reusable vehicle of the same gross mass by a factor of 10 compared with that using current materials. Thus the combined effect of the introduction of Buckytubes as well as the other new technologies could, in principle, reduce launch cost to LEO by a factor of up to 10,000 from current levels.

Concurrent introduction of propellantless tethered, electrically powered, or solar sail orbit transfer vehicles will extend these factors to GEO and beyond, for even greater reductions of launch costs to destinations beyond LEO. These factors, if additive, could make the launch costs of any spacecraft $1 to $10 per kilogram, which is essentially "free" compared with today's costs. This means that it will cost only $1000 to $10,000 to launch a 1000-kilogram spacecraft. However the spacecraft weights to be orbited will also have shrunk by factors of up to about 100 as discussed above, and spacecraft weighing more than a few hundred kilograms will be rare regardless of their size or function. Thus the costs to launch most spacecraft to essentially any orbit will usually be on the order of magnitude of a few thousand dollars, which is similar to current air cargo rates. While practical manufacturing limitations may prevent the superposition of these weight savings and the full attainment of these great reductions, even a partial attainment makes it clear that access to space will indeed be "essentially free," compared with the situation today.

18.5 Extremely Small and Extremely Large Space Systems

The current rush to "downsize" spacecraft, in many instances sacrificing performance to obtain lower costs, is incorrect by itself. This is because large space systems can also be very lightweight and inexpensive, and yet perform completely beyond the capabilities of small spacecraft acting alone if they adopt the technologies discussed in Part I. This not to say that the current trends to downsizing spacecraft are wrong, quite the contrary—in fact they have not been carried far enough. But the drive toward making spacecraft smaller today is principally to reduce costs because today it is unfortunately true that spacecraft weight equates to cost, both directly and for its launch.

However, when launch costs drop radically, the current emphasis on weight reduction will lose much of its rationale and urgency. Spacecraft weights will be further reduced by using lightweight but large membrane antennas and optics to develop very large spacecraft. In the future, therefore, very large spacecraft will also be lightweight and inexpensive, even if constructed from relatively conventional materials. In addition, the advent of new materials such as Buckytubes will reduce even further the weight of all components and subsystems in a spacecraft by up to two orders of magnitude. Spacecraft weight will then plummet so as to be truly insignificant. Thus large and small space systems alike will be inexpensive, with their design driven not by weight or cost but by best match to mission needs.

Additionally, adopting swarms of tiny but coherently cooperating spacecraft that can be truly "mass produced" to perform large tasks will also make other classes of large and highly capable space systems inexpensive and lightweight. The net result of the foregoing is that large space systems made up of swarms of small spacecraft will be no more expensive than larger spacecraft acting alone—they will both be "low cost." This leads to a complete departure from the current trend to downsize to save money. This message, shown in Fig. 18.1, may not be very popular with those generations reared in the virtues, necessities, and skilled art of weight reduction in all spacecraft components no matter what the cost, but it is an inescapable conclusion once both launch costs and inherent spacecraft weights drop radically.

Figure 18.2 shows the implications of greatly reduced weights and costs of both small and large space systems. Weight and cost reductions of systems in orbit that will occur solely because of introduction of new technologies, without Bucky-tubes, will be by factors of 100 to 10,000. The effect of using Buckytubes alone will be an additional effect of factors of up to 1000. If completely additive, in principle, the total effect of both the new techniques and technologies as well as the use of Buckytubes, taken together, could be reductions in effective weight and cost of up to 100,000 to 10 million from those common today. Even if practical manufacturing limitations result in effective factors no greater than 100–1000 for all effects combined, the net results would be far-reaching.

By any measure this is indeed a first class revolution!

18.6 Spacecraft Appearance

A linear extrapolation of present spacecraft techniques into the future would lead one to conclude that future spacecraft will either simply be larger buses with larger payloads or many independent small spacecraft weighing about 100 kilograms each. Such conclusions would be far from correct. Although some such spacecraft will undoubtedly exist, many more will actually look very different in 10 to 30 years or more.

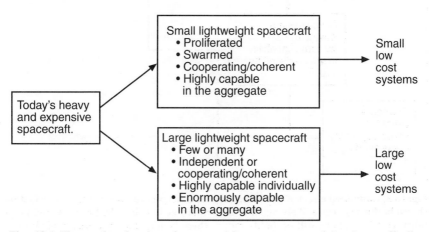

Fig. 18.1. The results when launch costs and large structure weights drop radically.

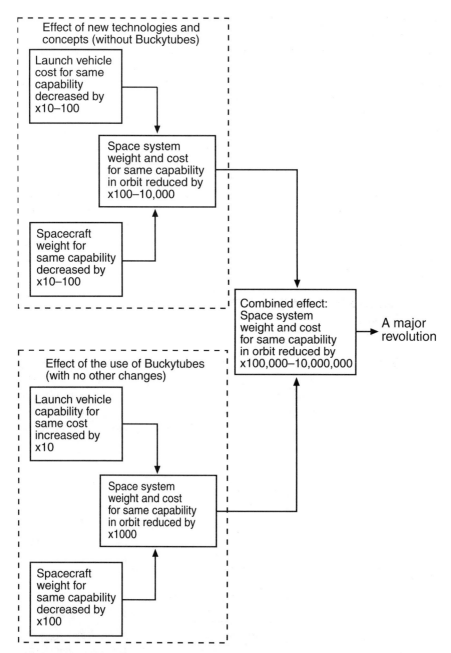

Fig. 18.2. Combined effect of new technologies, concepts, and Buckytubes. The factors shown in this figure are likely maximums, achievable in principle. Practical and manufacturing limitations will probably reduce these factors, but even then their impact will still be a major revolution

This is because a linear extrapolation of today's technologies does not lead to large adaptive membrane apertures, stationkept rather than attached to the rest of the components or with the components connected by tethers. Nor does it lead to swarms of hundreds of thousands of formation-flying picosatellites cooperating coherently under central control to form apertures or other new functions that do not exist without such cooperation.

Furthermore, linear extrapolation from the present continues the all-consuming practice of extreme weight reduction in spacecraft under the assumption that launch into space will continue to be hugely expensive, resulting inevitably either in expensive capable spacecraft or in tiny ones that perform only tiny jobs. The cost reduction advantages of mass production are never attained.

Thus linear extrapolation of current technologies into the future will inescapably lead to heavy, expensive, and not very powerful space systems, and will be completely wrong as linear extrapolations usually are over the long term. What will happen is the exact opposite, and specialists in weight reduction will be looking for other employment. This book has tried to show a few examples of what spacecraft may look like and what functions they could perform if full use were made of the enabling techniques in Chapter 2. Judging from the plethora of powerful space system concepts advanced in this book, let alone many others that surely can and will be thought about by others, spacecraft types will indeed be enormously different from what they are today, and their numbers will be far greater. It is not possible to get those results by simple extrapolation using projections of current technologies.

18.7 Spacecraft Servicing and Upgrade

Routine servicing and upgrading of spacecraft in orbit, let alone on the ground, has never made economic sense because of high launch costs, not even for constellations of several satellites in the same orbit. However, when launch costs drop by two to three orders of magnitude, the equation will change completely, as it will be economical to service and upgrade many spacecraft systems.

Routine servicing and upgrading could take place in orbit or on the ground. This will also apply to GEO and other orbit spacecraft as orbit transfer costs to access the higher orbits plummet. Routine and inexpensive accessibility to spacecraft for repair or replacement will also reduce the need to make them long lived, and cost savings will then also accrue in some elements and subsystems that need not be designed for very long unattended life, accounting for even more overall spacecraft cost reductions. In addition, the ability to upgrade will be increasingly valuable as technological obsolescence times drop to months or years as opposed to decades today, enabling space systems to attain almost continuous maximum performance and economic returns.

In fact the low cost and ease of access, and the ability to design and manufacture with servicing in mind, are traditional features of most Earth-based and airborne systems. Thus it is expected that with the introduction of satellite servicing, space operations will become no different and no more costly for many spacecraft than operations of ground or air systems.

18.8 Spacecraft Proliferation and Expendability

Not all spacecraft will be serviced. Many picosats that will make up the coherently cooperating swarms would be essentially flying single chips—highly integrated tiny spacecraft. Most of these, and many larger spacecraft, will be grown in microcircuit furnaces rather than assembled from separate components as are current spacecraft. When assembly is eliminated, unit costs drop in volume production, and it becomes more economical to expend and replace the units than to service them.

The concept of expendability will also apply to large gossamer membrane sensing and communications spacecraft because the capture and recovery, let alone the handling and refolding, of these systems is neither practical nor necessary. Thus the mix of spacecraft in the future will include some large and serviced, some tiny and expended, and some large and expended. Each will be far less expensive than current systems to develop, manufacture, launch, deploy, and operate.

Some will undoubtedly raise the spectre of massive amounts of uncontrollable orbital debris as an argument against these trends. In fact, the same technologies will give rise to versatile and inexpensive specialized spacecraft permanently stationed in many orbits, whose function it is to rendezvous with the larger debris and shepherd them into safe disposal orbits or to deorbit them. In addition, NASA studies[18.1] have shown conclusively that essentially all orbital debris smaller than a fraction of a meter across can be deorbited rapidly for reasonable cost with nearly current technologies, and NASA even performed preliminary design definition of such systems. Thus there is little reason to fear that the sky will be black with orbital debris that would prevent safe operation of space systems.

18.9 Brassboards and Spares

Frequent and inexpensive space access will make the weight of spacecraft essentially irrelevant. This is completely different today, where weight reduction is both "king" and "holy grail," and is mostly responsible for the high cost of spacecraft.

In contrast, routine, easy, and inexpensive space access will also make routine the orbiting of "brassboard" spacecraft and components, or basically developmental units in various stages of readiness. Testing spacecraft elements and assemblies in their expected environment will be a routine step in the development process, as it is for ground and airborne systems today. The actual operational spacecraft will be much more ready to be used in space after final developmental steps and tests are carried out in space. The total costs, including the cost of failures, will be far smaller and the spacecraft will prove far more reliable with such space developmental testing.

Because spacecraft can be fielded at various stages of development and because all spacecraft will be low cost, whether final operational units or development and test items, fielding many spare operational spacecraft will become common practice and economically viable. Thus switching to orbiting spares while the failed spacecraft is either repaired or disposed of safely will be routine, with operations continuing uninterrupted.

18.10 Megawatt Power Levels

The high efficiency solar energy conversion, high capacity and efficiency thermal dissipation, and the gossamer aperture concentrator technologies of Chapter 2 will result in the development and fielding of megawatt level power modules that could cost as little as $20 million and weigh as little as 2000 kilograms, when constructed from conventional materials. Their weight would be reduced to 20 kilograms and their cost would approach $200,000 if constructed entirely out of Buckytubes. Production units of these power modules will be developed by the private sector as a part of commercial power delivery initiatives, although the government's traditional role of supporting research and technology risk reduction will be necessary to enable such commercial investment. Once completed, these modules will be available to the military and civil space programs at only marginal recurring unit costs.

Low cost and lightweight megawatt level power modules combined with robotic assembly techniques, high power RF technologies, large gossamer antenna technologies, and mass production will bring about space beam power delivery systems whose significantly lower development and operating costs will result in delivered electricity costs competitive with those of nuclear and fossil fueled power plants, but with no deleterious effect on the environment. As a result, commercial power delivery satellites at the gigawatt level will become widespread, although only starting in the latter portions of the time frame considered here.

The availability of the megawatt level power modules at low weight and cost will also enable high power radar and laser sensing as well as force projection from space, and a plethora of other high power applications not otherwise possible or economically feasible.

18.11 Powerful Information Transmission and Processing

Many of the functions described in the advanced application concepts of Part III are dependent on the ability to move, process, and store vast amounts of information from multiple sources. The transmission needs will eventually be met by laser links to, from, and within space, using path redundancy to circumvent weather effects. Processing and storage needs will be met by the growth in capability and decrease in cost and weight forecast for computers for the foreseeable future—the progression of Moore's law (based on Gordon Moore's 1965 observation that circuit densities of a semiconductor would continue to double regularly, which they did annually until approximately 1974 and every 18 months after that).

There appears to be no fixed end to this law. While current technologies such as optical lithography of circuits in chip manufacture may reach their limits in the near future, the introduction of other technologies such as biological systems, Buckytube semiconductors, or quantum effects could reverse this slowdown and even greatly accelerate growth. In fact doubling speed or capacity roughly every two years in the information field had been happening for many decades prior to Moore's observation, but driven by different technologies, and that trend will continue without foreseeable end.

The most confident statement that this book makes about future capabilities is that information processing, storage, and transmission power required during the next 30 year time frame for any of the advanced space system concepts identified in this book will be available when needed. The commercial markets will drive this proliferation, and the government will fund the basic research.

18.12 Highest Leverage Technologies

The highest leverage technologies that should be developed to enable the advent of the advanced concepts of Part III are summarized in Table 18.2. This list is not intended to be in any priority order. Furthermore other technologies that should also be developed were not treated in this book due to resource limitations, but their absence should not be interpreted as intentionally exclusionary.

Table 18.2. Highest Leverage Technologies

- Adaptive piezoelectric reflector membranes, actuated by electron beams
- Coherent cooperating distributed or swarmed spacecraft of all sizes
- Buckytube matrixless and composite structures and spacecraft components
- Long lightweight, high strength long-life tethers, wire and nonconducting
- MEMS FEEP integrated micropropulsion assemblies
- Formation flying techniques with submillimeter relative position accuracies
- Spectrally split, multiple matched bandgap cells in concentrated solar power arrays
- Liquid crystal spatial light modulators with more than 1 millimeter of time delay correction
- Micro-particle stream heat radiators
- High capacity information transmission, processing, and storage to meet all needs

18.13 References

[18.1.] Jon Campbell, editor, "Project Orion: Orbital Debris Removal Using Ground Based Sensors and Lasers," NASA Technical Memorandum No. 10852 (Marshall Space Flight Center, October 1996).

19 Perspectives

This chapter discusses a number of general perspectives on the applications of Part III and the observations of Chapter 18. These are listed in Table 19.1

19.1 Space Functions Will Migrate to GEO and GSO

The majority of noncommunications space applications today are in low Earth orbits. This is not because of some inherent advantage that spacecraft have in low orbits, but rather principally because of economics, that is the high cost of access to space, the even higher cost to access higher orbits, and the high costs and demanding technologies of sensor and antenna apertures large enough to perform as well from higher altitudes as from low orbits. Penalties for low orbit use, however, arise because spacecraft cover a limited area and they move with respect to the surface. As a result, many spacecraft are needed to obtain desired coverage, operations are complex and expensive, and in many cases spacecraft have limited or no utility for a significant portion of their orbits.

The high costs of space systems and their emplacement will drop dramatically when the cost of access to any orbit is reduced by two to four orders of magnitude and the cost to develop and produce larger spacecraft are reduced by another two to three orders of magnitude. Then it will be no more expensive to operate from GEO, GSO, or even higher orbits than from LEO. In addition, only a few spacecraft will be required rather than many, and the system performance will be as great as that of many smaller spacecraft operating from LEO. As a result, many space functions will migrate to GEO and GSO, which will be the preferred orbits for many if not most space applications.

The oft-repeated, so-called "crowding" of these orbits, cited as a deterrent to their use, is a red herring. The scarcity of orbital slots in GEO for communication satellites stems from the limited (currently 2 degrees) beamwidth of the ground terminal antennas, which are fixed and do not track spacecraft, thus limiting the number of communication spacecraft in GEO to 180. However, even then there are 1400 kilometers between adjacent spacecraft—hardly a crowded situation. Furthermore, the use of large spacecraft antennas with many narrow beams,

Table 19.1. Perspectives from Assessing Applications and Observations

• Many space functions will migrate to GEO and GSO for coverage and low cost.
• Most functions that can now be done from aircraft will be done from space.
• Many global functions will be done remotely from the haven of CONUS.
• Many space functions will remove air and ground crews from harm's way.
• Commercial activities other than communications will dominate space.
• It is better to create wealth than to manage scarcity.

larger ground terminal antennas, tracking capability, higher frequencies, or all of these will lead to a greatly expanded capacity of the GEO arc to support communication satellites without physical crowding.

Notwithstanding the above, relief from perceived crowding of the GEO arc is possible and feasible. All that is required are ground terminal antennas that are capable of very slow (24-hour cycle) two-axis tracking, rather than being fixed and pointing in one direction, which can surely become economically feasible in this time period given the hundreds of millions of customer terminals that will be in common use. Then a large number of spacecraft can be placed either in inclined geosynchronous but not geostationary (GSO) orbits, whose subsatellite points describe figure-8 patterns on the ground, or in "halo" type orbits that appear to circulate about an equatorial point. The subsatellite points of these latter orbits describe circles projected on the ground.

Large numbers of spacecraft can be placed in such orbits without interfering with each other, and the desired spacecraft tracked by ground terminals. Many such GSO orbits could be implemented around the globe, vastly increasing the total communication satellite capacity. Since the halo orbit spacecraft have eccentricity as well as inclination but still have the same 24-hour period, the subsatellite tracks of a number of these orbits could be elliptical or circular and concentric without any close approaches or collisions in space because the ascending and descending nodes occupy different altitudes. Figure 19.1 illustrates four such orbits centered on a longitude. The circles represent spacecraft with 2-degree spacings to each other. These circulating orbits hold 200 spacecraft while only "occupying" one longitudinal GEO "slot." Thus it should be possible to place 10–100 times more spacecraft in 24-hour synchronous orbits than the 180 satellite capacity limit of the geostationary arc, putting to bed the issue of perceived crowding.

Notwithstanding the above discussion, both the limits of the GEO arc and the potential GSO solutions apply only to communication satellites. It must be emphasized that for sensing and other noncommunication spacecraft, there is no such 2-degree spacing limit and there are no "orbital slots," and therefore no such limits exist. Even the 1400 kilometer distance between communication satellites in GEO is hardly an impediment for the use of GEO by hundreds if not thousands of other spacecraft.

Orbital debris is an additional positive factor in using these high orbits. In GEO or GSO the relative velocities of potential debris collisions are much lower than in LEO and can be more easily avoided by small altitude separations, as the eccentricities of spacecraft in these orbits are generally much lower than of those in LEO. Thus the formation of secondary debris by collisions between spacecraft can be more easily avoided, and the consequences of actual impacts will be lower. The one disadvantage is that all orbital debris in such high orbits are very long lived as the orbits are inherently more stable in the presence of solar and lunar perturbations and there is no natural atmospheric reentry filtering; thus if there is a problem it must be solved by active means.

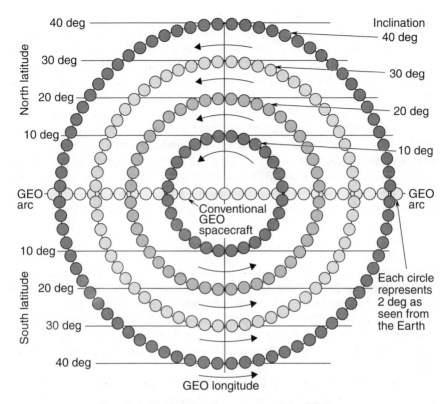

Fig. 19.1. Multiplying the capacity of the GEO arc.

19.2 Functions From Space

A number of military and some civilian applications and functions are migrating from ground to airborne use to gain the advantage of range and visibility, particularly in many battlefield and some civil sensing applications. Such applications are usually power and/or aperture limited, as the requirements for both generally grow with the square of the distance for passive sensors and with the fourth power for active sensors. Thus while aircraft do obtain considerably greater range of operations because of their altitude, they require higher power, larger apertures, and entail higher operation costs than ground devices.

   Not only are direct aircraft operating costs very high, the indirect or infrastructure costs of maintaining the airfield, repair facilities, refueling facilities, and the personnel to operate them are usually not charged against their direct operations costs. If such infrastructure costs were added, as they should be in any fair comparison, the true aircraft operation costs would be even higher.

   In the future, when large apertures and high powers can be implemented in space and enabled by technologies of Chapter 4, the infrastructure and operations costs of space systems will be very much lower. This will be so particularly if GEO orbits can be used, as only one spacecraft will be required to have continuous visibility over any theater or region, and only a few will be required for global coverage. Large numbers of aircraft would be required to do the same job, both because of the smaller visibility from any one aircraft and the large and probably airborne logistics tail involved for support of such aircraft operations.

Thus replacement of airborne operations by space operations will generally result in more effective applications and operations at lower cost. This is not true today, of course, but will become so when the identified technologies are available to support the functions from space. Space will not totally replace air and ground media for applications, of course, but will be used where advantageous, resulting in a mix of ground, air, and space systems for both observation and force application weapon functions, each chosen for maximum effectiveness at lowest cost.

19.3 Global Functions from CONUS

Space reflectors in many precise shapes, enabled by the adaptive gossamer technologies of Chapter 4, will usher in an era where many military functions of sensing and force application will be performed from the sanctuary of rearward areas and even from the CONUS. The current "denied areas" will greatly shrink if not disappear, and the forward force structures will be much smaller without losing effectiveness. Much greater effectiveness and greatly reduced operations costs will result.

Operating these systems from CONUS rather than from overseas locations will greatly increase security at much lower cost, particularly if the systems would otherwise be required to be close to theaters. The CONUS-based operation of these systems will also be readily serviced and upgraded at lower cost than maintaining forward based systems, and the added burden of the logistics train otherwise required is avoided. Thus security, safety, and effectiveness of functions performed will increase, while costs, force structure, personnel, logistics, and denied area sizes will decrease.

19.4 Removing Crews from Harm's Way

Much as the introduction of unpiloted, combat air vehicle airborne elements will reduce the danger to pilots and other fighting personnel in this time frame, the introduction of space-based systems and CONUS-based remotely operated systems will result in further removing crews from harm's way. This will be particularly evident as space begins to be used for force application in addition to today's sensing and communication functions that are principally those of force support. The decreased numbers of people at risk, both in the air and on the ground, will greatly reduce the number of casualties, and will thus remove a powerful barrier to aggressive and effective military operations that exists today.

The importance of reducing potential casualties resulting from any contemplated or real conflict cannot be underestimated, as it is intimately tied to the political leadership's willingness to undertake military operations, as well as to the public's willingness to support them. The use of space, particularly for remote control that reduces the number of crews in harm's way, can completely change this picture by drastically reducing the number of expected casualties.

19.5 Commercial Activities

The low costs of development, launch, and operation of space systems in this future time period, together with their many potentially highly lucrative applications, will result in a virtual explosion of the commercial uses of space. These commercial uses will completely dwarf today's commercial communications satellite market. The magnitude and scope of many future commercial programs will greatly exceed those of defense and NASA uses of space, both in numbers and sizes of spacecraft orbited and in the moneys devoted to them. While this is the reverse of the situation that exists today, it is completely analogous to the bulk of nonspace commercial activities in free market economies. Of course, lower space access and spacecraft development costs will also greatly increase the defense and civil agency uses of space in this same time period, but nonetheless new types of commercial activities will dominate the uses of space because of their very large profit potential.

Expanded and highly capable commercial space activities will, of course, develop in communications and observation. However the preponderance of commercial activity will exist in entirely new areas including the generation and delivery of electrical power and light to space and ground users; the development and operation of many enterprises together constituting an orbiting infrastructure for manufacturing, entertainment, tourism, package delivery, fast people transport, and sports, as well as the commercial development and operation of transportation vehicles for access to and from space and the moon.

While these facilities will be developed principally by and for the commercial sector, some of their capabilities will undoubtedly be leased or otherwise utilized to a significant extent by the defense community. The adoption of commercial activities for defense purposes is well under way today, and will only increase in scope and magnitude as it will be increasingly less expensive for the Department of Defense to lease commercial capabilities than to develop its own. The increasing capability of commercial space systems will make this trend unstoppable. There will always be a need for strictly military systems, both for the United States and for its allies. Nonetheless the sheer magnitude of commercial space activities will make them a critical aspect of national security whether they directly support defense uses or not because they will represent critical national infrastructure.

19.6 Creating Wealth

Throughout history the world seems to have been dominated by those who try to maintain or slightly improve a known capability and stifle processes that would foster innovation. Unfortunately, too often this is motivated as much or more by the desire to maintain bureaucracies and power structures as by honest attempts to achieve a real increase in capabilities at modest cost. Whatever the cause, such attitudes and actions limit options and preclude achievement of the very developments that would bring about greatly enhanced capabilities at radically reduced costs.

This book advocates the exact opposite: Investing resources on "deliberately disruptive" technologies and new ideas will simultaneously create new and far more effective capabilities, greatly expand market horizons, and ultimately reduce resource requirements. This is as true for military as for civil and commercial activities, as many new capabilities in each are created with fewer resources. In short, more is achieved not by contracting the vision in order to be frugal, but by creating entirely new options and directions.

The net result is that wealth is created, both in dollars and in capabilities, which is a far more desirable outcome than merely stretching resources—the management of scarcity. This book is dedicated to illustrating that principle by example.

20 Implications for Defense, NASA, and Commerce

The previous chapters described a number of new technologies that will be attainable in the 2000–2020 time frame and presented many space system applications that these technologies could enable in the 2010–2030+ time frame. These applications demonstrate the power, diversity, and utility of the future uses of space. Though merely examples, the concepts amply demonstrate that many powerful and useful, if not revolutionary, defense, civilian, and commercial space applications could become realities in this time period.

Rationales were also developed as to why weights and costs of small as well as large spacecraft will drop by a factor of up to 10–100 and why the costs of orbiting them will drop by factors of up to 100–1000, so that the costs of building and orbiting the spacecraft could probably be reduced by factors of up to 1000–100,000, though their full realization may be limited by practical manufacturing considerations. Large as these reduction factors are, they are attainable without resort to exotic new materials. However, when Buckytubes are introduced into both spacecraft and launch vehicle construction, in principle, additional reductions by factors of 100–1000 will result. As discussed in Chapters 9 and 18, practical considerations may not allow the simple superposition or combination of the effects of new technologies as well as the introduction of Buckytube materials, which, in principle, could result in total weight and cost reductions of factors of 100,000 or more from today's levels. Nonetheless, the reductions will certainly be huge, making the development and fielding of space systems no more expensive than that of ground or airborne systems, yet producing systems of enormous power and utility.

Weight, which is today the major determinant of space system cost, will become essentially immaterial in the future. Weight and cost will be very low even for large aperture spacecraft because of their gossamer, structureless construction, as well as the use of Buckytubes for most of their components. The use of Buckytubes will also assure low weight for structure-intensive spacecraft such as crewed systems and large space construction infrastructure. The time required for development of space systems will be greatly reduced by avoiding much of the labor-intensive design, development, and testing required to make spacecraft simultaneously lightweight and reliable in long term unattended operations. Mass production techniques adapted from Earth-based manufacturing practices for the swarms of picosats and the highly modular larger power systems will further lower production costs. The net effect of the great cost and time reductions will be that the use of space will grow to such high levels of activity and to encompass so many new functions as to be difficult to comprehend today. The implications are far reaching for defense, civilian, and commercial future uses of space, which will be fundamentally different from those of today.

20.1 Implications for Defense Use of Space

The new capabilities and radically reduced costs of space systems will have profound implications for national defense. These are outlined in Table 20.1. One of the biggest of these changes will be that while today space is used largely to support the warfighter and is seen principally as a force enhancer, with the advanced concepts presented here force will be projected directly from space. These force projection functions will include jamming and other electronic countermeasures, information warfare, precision physical delivery of weapons, delivery of high energies, and more. These systems will continuously project force with precision, timeliness, surprise, coverage, and in such quantities so as to be devastatingly effective in tactical engagements in any theater anywhere. Force projection from space will be so effective and relatively inexpensive compared with other options that space force will be used frequently, chosen over force applied from the air, sea, or ground.

The role of space as force augmenter will also be greatly expanded, providing the long-desired complete situational awareness—an unprecedented capability for continuous cognizance of large and small scale strategic and tactical situations. The ability to detect, classify, track, and target ground, air, and space targets continuously anywhere on the globe will result in more effective engagements with fewer personnel and equipment, and at greatly reduced cost.

Table 20.1. A Whole New Ballgame—Defense Space

- Global force projection from space will be ubiquitous and devastatingly effective.
- Complete situational awareness will exist from GEO at theater to global scales.
- Many crews will be removed from harm's way by performing functions from CONUS.
- Precision weapons will be delivered globally from CONUS.
- The size of, and need for, logistic tails to support operations will be greatly reduced.
- Space radar will mostly replace AWACS, JSTARS, SAR, and SPACETRACK.
- Spacecraft development, deployment, and operations costs will approach those of aircraft.
- Some space systems will be incrementally funded, emplaced, and upgraded.
- Most of the advanced ideas of SAB's "New World Vistas" will be fielded.

BUT

- The United States will not have decisive technological advantage over others.
- Commercial infrastructure and services will dominate space activity.
- Congress will insist that DOD use these capabilities.
- We will have to learn to observe, fight, and win in this environment.

Just as important is the great reduction in the number of crew and support personnel placed in harm's way by the use of space systems as compared with conventional engagements, leading to far fewer casualties. A key element of this will be the ability to perform many functions directly from CONUS, reducing the need to place people in forward land, sea, or air positions, and accomplishing the saving of lives. The result will be a combination of remotely presented presence and on-the-scene presence globally, achieving the Air Force's global presence mandate.

People-intensive operations such as flying AWACS and JSTARS aircraft, key and expensive elements of today's tactical battle support operations, will be replaced by multifunctional space-based radar systems that will be more capable yet less costly to operate, especially when all actual support costs are included.

The increased effectiveness of tactical and strategic operations engendered by these space capabilities will greatly reduce the weight and fuel requirements of most materiel, making major reductions in the logistics "tail" needed to support military operations.

Because of the low cost of both spacecraft and launch vehicles to accomplish these functions, the same or even reduced budgets will suffice to acquire many more space systems, reducing resistance to their introduction and encouraging their proliferation. Furthermore, many new space systems will be developed and emplaced incrementally requiring only incremental funding, thus making their fielding and upgrading more readily affordable in a realistic budget environment. Thus the new technologies and techniques will enable DOD to accomplish much more with smaller budgets by using space for many of the functions.

These new technologies will allow the actual implementation and fielding of many of the ambitious advanced space system concepts identified in the Air Force Scientific Advisory Board's "New World Vistas" study, the U.S. Space Command "Space Vision 2020," and other forward looking planning studies, which identify many desirable capabilities in space but only hint at how to actually attain many of them.

The net result of the above capability projections is that in the next 10–30+ years, space systems will become an absolutely indispensable element of the defense force structure for force projection as well as for intelligence, surveillance, and reconnaissance. The nature, magnitude, and effectiveness of these space capabilities will be well known to friends and potential foes, and that knowledge itself will surely preclude some hostilities. If and when deterrence fails, these same space capabilities will allow effective and rapid pursuit of hostilities and hasten their favorable conclusion, with fewer casualties and at lower cost.

A downside to these new capabilities is that the enabling technologies will, in all likelihood, be available to all nations and groups including the so-called rogue states and some terrorist organizations, given the rapid globalization of information systems. Thus the United States will be unable to use technological advantages in space to the extent that it did in the Gulf War, and will likely confront many of these advanced space applications fielded by hostile powers or groups. In order to prevail in this situation, the nation must place greater reliance on sound strategy and tactics, rapid system development and deployment, and faster and further technological innovation in order to exploit the potential of space. It must also learn to think "asymmetrically," as the capabilities we will face may be more rudimentary yet very effective in supporting these organizations' goals.

Thus it is very important that the United States develop these space application systems aggressively because if it doesn't, and other nations or groups do, they will have attained the high ground and placed this country at a tremendous disadvantage. In fact once this high ground is unilaterally attained by anyone, leveling the playing field will be much more difficult because the party with the advantage of some of the illustrated space capabilities could well prevent anyone else from successfully orbiting similar systems or even any systems. It is thus vital that the United States seize the initiative and capture this high ground.

In addition to purely defense applications, a broad array of commercial applications, far beyond today's mature communications and limited Earth observation systems, will eventually dominate space activities and will be available for DOD utilization. In time, using fully capable commercial systems will be much less expensive and faster than developing similar capabilities exclusively for government use. Though these trends are already evident today, there will be an inexorable acceleration of the trends. Inevitably DOD will be increasingly directed to use these capabilities by the Congress when setting the bounds and budgets for DOD space programs. This may save money while introducing some new vulnerabilities. Thus the distinction between military and commercial systems that is already beginning to blur will accelerate and broaden. This may save money while introducing some new vulnerabilities.

The United States must learn to observe, communicate, fight, and win in the fundamentally different environment from that of the past, which the new space technologies and systems will create.

20.2 Implications for NASA and Civil Agencies

The implications of the new technologies and the space system capabilities that they will engender are no less profound for NASA and other civilian government applications than they are for defense uses. These implications are outlined in Table 20.2.

Using the technological advances of this report, NASA will be able to attain affordably and rapidly some of the "grand challenges" for which it has long strived. The advent of advanced in-space propulsion technologies including spinning tethers, inertial confinement fusion, and beamed energy, combined with two to three orders of magnitude effective weight reduction resulting from using Buckytubes to construct both spacecraft and launch vehicles, will allow travel to many bodies within the solar system in months instead of years or decades, greatly shortening the time to obtain science return from planetary missions.

Major benefits will accrue to crewed missions to Mars. In addition to greatly lowering mission costs and thus increasing their likelihood of being approved, use

Table 20.2. A Whole New Ballgame: NASA and Civilian Government Space

- Exploration
  - Traveling anywhere within the solar system in weeks, not years
  - Avoiding all deleterious effects on humans of flight in zero gravity
  - Developing and utilizing extraterrestrial resources

- Astronomy science
  - High resolution imaging of Kuiper belt, Oort cloud objects, and planets around other stars
  - Enlarging the observable universe volume by a factor of 1000

- Earth science
  - Developing comprehensive microwave and optical characterization of Earth's surface
  - Comprehensive understanding of weather and climate worldwide
  - Improving the ability to predict earthquakes and floods

- Commercial space development
  - Enabling the commercial provision of abundant, inexpensive, clean energy to Earth
  - Enabling major new commercial industries such as space tourism

- Humanity
  - Protecting Earth from threatening asteroids and comets

of the technologies will avoid the degradation of the human body in the zero gravity environment. A reduction of these effects will be due simply to the shorter trip duration. A complete avoidance of all such effects will be made possible by producing artificial gravity in the Mars spacecraft in which the habitats are separated from other massive components by a short (one to a few kilometers) fail-safe tether rotating at less than 1 revolution per minute, thus providing any desired gravity level in transit without sensible Coriolis force or gravity gradient disturbances to the crew. This tethered rotating configuration will also be adopted for space stations that are to function as transportation nodes in orbit around the Earth, moon, or Mars or in Lagrangian points.

The advent of advanced propulsion, tethered rotating spacecraft, and extensive use of Buckytubes will usher in an era of affordable human planetary exploration in which mission costs now projected to be in the tens to hundreds of billions of dollars will be reduced to at most tens to hundreds of millions dollars. This new affordability will ensure that large numbers of human missions to the moon, Mars, or other solar system bodies will be approved and undertaken, and that their science and public relations returns will be forthcoming relatively quickly.

The low cost of missions and operations beyond Earth enabled by the advanced technologies and use of Buckytubes will finally allow at least the beginnings of permanent outposts, if not bases, to be installed on the moon and Mars in this time period. Extensive processing and use of local in-situ materials will become routine, both for use in the bases themselves and for export to Earth orbit. Many of these facilities, as well as spacecraft to generate and deliver power to them and to the transfer vehicles supplying them, will be commercially operated, with the government a major but not exclusive customer and will be enabled by government developed technologies. This is discussed further in Section 20.3.

NASA will be able to afford to implement the very large aperture and high resolution telescopes required for detailed imaging of planets around other stars, image outer solar system objects such as those in the Kuiper belt or Oort cloud, and perform astronomy and astrometry with orders of magnitude higher resolution and sensitivity than either the Hubble Space Telescope or its successor, the James Webb Space Telescope (JWST). Though some of these systems are being defined today, their costs are projected to be so high as to make them either unaffordable or relegated to be approved late and with stretched schedules, which will greatly increase their already very large total costs.

However, with the advent of adaptive membrane telescopes with stationkept elements and no truss structure, these systems become a factor of up to 10–100 times lighter and less expensive, yet having even larger apertures and separations than contemplated today. The additional use of Buckytubes throughout to manufacture the spacecraft and their launch vehicles will reduce costs by another factor of 100–1000, and the combined effect could in principle reduce costs by a factor of 1000–100,000 or even more in the long term. While practical manufacturing considerations may not allow the full superposition of these reductions, compared with today's levels the low resulting costs will make these systems completely affordable and likely to be approved with little opposition.

These instruments will have apertures in the 25–250-meter-diameter range with 10 to 1000 times more sensitivity than JWST, enlarging the size of the observable universe by a factor of 3–30 and making possible observations to the very beginning of time. In fact, with a properly funded and prioritized technology program for adaptively corrected piezoelectric bimorph membrane optics, stationkept elements, and large corrective range liquid crystal modulators, a much larger telescope than JWST could be developed which would be far lighter and less expensive, and yet deployed only a few years later.

Earth observation systems will also take advantage of the higher spatial and spectral resolution that will be available, the continuous or long dwell capability, and the ability to image in the microwave spectral region with high spatial resolution as well. The continuous dwell and large apertures will allow synoptic whole Earth observations, as most systems will be deployed to GEO or GSO altitudes. Some of these space systems will be composed only of large orbiting reflectors, allowing the complex and more expensive detection equipment to be left on the ground while accessing many or most places on the globe by viewing them via reflections from the space mirrors.

These system capabilities will result in much more complete characterization of the Earth's surface than current Earth observing systems or their contemplated successors, and at orders of magnitude lower cost. These systems will also provide much more complete characterization of the atmosphere, yielding a solid understanding of the processes influencing climate, weather, pollution, and environmental adverse effects. These observing systems, together with wholesale use of expendable very low cost micro-sonobuoys, micro-weather balloons, micro-ground sensors, permanently tethered high altitude weather balloons, and laser retroreflectors for measuring surface water levels will result in much better understanding and predictions of climate, weather, and hydrology, and aid in flood prediction efforts at reduced cost and schedules.

The extremely low cost of these space systems will allow them to be rapidly fielded to mount nearly continuous scientific campaigns. The number of space systems that will be approved as well as their ambitious goals and schedules are difficult to imagine in today's austere budget climate, and will shorten by decades the time required to gain understanding of the targeted phenomena.

The ability to inexpensively instrument most earthquake faults and other geological features on the planet, and to read them out essentially continuously with great accuracy at extremely low operations cost, will greatly aid the ability to predict at least some earthquakes and other natural disasters with useful warning times. It will not take very many such predictions, with their probable savings in lives and property, for NASA to become heroes again and therefore to see expanded budgets as well as public approval.

Nonetheless, as appealing as some of these capabilities in science and exploration may be for NASA today, they do not address its decades-long preoccupation with science and exploration to the near exclusion of support for commercial space activities.

NASA has a statutory obligation to promote the economic well-being of the nation by enabling and supporting commercial activities in space that would create national wealth and enhance national prestige. Although NASA adheres to that role in aeronautics, as it has done since the days before 1959 when it was the National Advisory Committee for Aeronautics, it has long neglected the enabling of major future commercial activities in space as an aspect of national security, concentrating on science, exploration of the solar system, and a human spaceflight program driven by astronauts. While no other government agency can or should undertake space exploration programs, NASA must step up to the development and demonstration of specific technologies needed by commercial industries if it is to undertake major longer term commercial space capabilities because it does not have the private funds to afford such development. This technology development role in support of industry is a classic role of government to enhance the economy and national security over the longer term.

Thus the government especially through NASA needs to develop and place on the shelf technologies that will enable a vibrant space tourism industry, wholesale solar power delivery from space to Earth, space business parks and hotels, infrastructure to support public and private operations on the moon, asteroids, and Mars, as well as a host of commercial programs that were identified in Part III. The government should not, and cannot, undertake the application programs themselves—that is a private sector function.

NASA does support advancing the technology for lower cost access to space, but at too slow a pace and with a risk-averse approach that effectively precludes anything beyond evolutionary growth, and allocates pitifully small funding for many enormously promising concepts that are beyond the immediate next generation of vehicles. Thus the acquisition of revolutionary capabilities for future access to space, which could operate at several orders of magnitude cost reduction with equal magnitude reliability and safety increases, is pushed decades into the future by delaying development of their enabling technologies.

One final, but extremely important, civil need is the continuous surveillance of space to detect Earth-threatening asteroids and comets, which would enable the early interception of the larger ones and prevent their impact with Earth. This threat is real with potential effects ranging from devastating regional destruction to extinguishing essentially all life on the planet. This threat has not been taken seriously enough, except in the movies. Although modest programs using ground astronomical telescope programs are in place to detect some of the potentially threatening asteroids, they are inadequate to detect any but the largest, will detect none of the long-period comets that pose as great a danger to humanity, and do not address any interception actions needed if such objects are indeed discovered.

The ability to detect faint asteroids and comets sufficiently far away so as to be able to mount an interception defense requires enhancing ground astronomical telescopes initially with a 1-meter-class instrument stationed at the L2 point and subsequently with several 25-meter aperture class space-based telescopes. In order for these to be affordable they will require the space imaging sensor technologies presented in Part I. Timely detection will allow launch vehicle-based interceptor vehicles to deflect or fragment such unwelcome celestial intruders. While some such interceptor vehicles could be assembled from current launch vehicles and weapon stores, the technologies of Part I will make the fielding of such interception systems far less expensive and enormously more capable than possible today, and with the capability to prevent cataclysmic damage to Earth, either in national or multinational programs.

While the detection of Earth threatening asteroids and comets may be a civil space program function, the mounting of an interception defense is probably a more natural DOD function. Regardless of how the responsibilities are divided, the new technologies, and the launch and space vehicles developed from their adoption, will be able to save Earth from enormous calamities and billions of casualties, if not the near-total extinction of life. Surely this is a worthwhile goal for the space program. It will be attainable with the technologies discussed herein.

20.3 Implications for Commercial Uses of Space

The current rapid increase in uses of space for Earth observation, communications, and related applications such as navigation will mature, but these applications will be completely dwarfed by the explosion in other, nontraditional, commercial uses of space that will likely occur in the 2010–2030+ time frame. Some of these commercial uses of space are listed in Table 20.3. Advanced low cost transportation to space will enable a near term rapid Federal Express-like service to deliver packages, and eventually people, anywhere on the globe within 45 minutes. This service could even begin with today's suborbital vehicles and be economically viable, but it will become economically highly attractive with low cost access to space, which will enable growth to large scale businesses and obtain the benefits of scale as the size of the market expands.

The availability of low cost orbital space transportation will enable many private firms to establish large scale space tourism operations in the 2010–2030 time frame. Current market surveys indicate that more than a million people annually would want to go into space if ticket prices of $10,000–$20,000 were offered, which could certainly be done with launch costs of $200 per kilogram. Even

Table 20.3. A Whole New Ballgame: Commercial Space

- Global rapid delivery of packages and people
- Public space travel (space tourism)
- Space hotels and cruise ships
- Power delivery to space stations and lunar and planetary bases
- Large scale solar power delivery to Earth
- Orbital sports pavilions
- Movie and advertising studios
- Provision of nighttime illumination to Earth
- Space business parks
- Production of commodities from lunar, martian, and asteroid resources
- Provision of the majority of launch and support services for DOD and NASA programs

- These and other commercial activities will be driven by the promise of lucrative profits.
- The magnitude and number of these programs will dwarf those of government programs.

if prices charged were as high as $100,000 per ticket, well over 100,000 people annually want to go into space. This would be achievable at $2,000 per kilogram launch costs. Total revenues expected from such ventures would be at least $20 billion annually, and economically viable businesses will certainly be established with that size lure. In fact, the surveys also indicate that 1000 people would go into space annually even if ticket prices were as high as $1 million, and more than 100 people worldwide if the prices were about $10 million. This is certainly borne out by actual data since current space tourists Dennis Tito, Mark Shuttleworth, and a Japanese journalist have paid, or will pay, nearly $20 million each for the trip. Space tourism has the potential to be at least 5–10 percent of the size of airline travel and tourism today, which is the second largest industry on the planet at $400 billion annually. Assuming new launch vehicles designed to orbit 100 passengers per launch, there will be at least 10,000 launches per year to support this tourism market, which is a launch rate two orders of magnitude greater than today's global launch rate.

Whereas the lowest tourism figures could be met by vehicles like the Space Shuttle in which passengers live while in space, development of space hotels as destination facilities will be needed to attain the larger number of space tourists. Thus a number of "Space Hiltons" or other hotels, small and large, will be developed and orbited by the private sector to meet this market. Probably starting small, by the 2030 time period these will become full-scale self-contained destination resorts, and eventually will evolve into facilities like luxury cruise ships, providing high levels of service to their orbital tourist clientele. This is not a question of if, but only of when. Orbiting these facilities and then maintaining them will by itself create a launch market greater than that of all space activities today.

Electrical power will be relayed by space reflectors from energy-rich areas of the globe to energy-poor areas, and especially from places where power from low pollution sources, such as hydroelectric dams, can be available. These power relay systems will be deployed and operating well before large scale power generation in space and its beaming to Earth become feasible. While limited operational systems for power beaming will begin to appear toward the end of the 2020–2030 time period, the full deployment of the hundreds of 10 gigawatt spacecraft that will eventually be needed will probably materialize somewhat later. Nonetheless, due to their very large size, weight, and complexity, these systems will create a large market for commercial launch services even in the time frame considered, second only to space tourism.

Other space facilities, less studied though clearly feasible and driven by the same market forces, will be deployed and operated by the private sector. These facilities will include orbiting sports pavilions where novel zero gravity games in which players float, bounce off walls, and fly in curved trajectories inside large sealed volumes will be televised to huge TV audiences worldwide. This could lead to "Cosmic Cup" soccer matches or the like, and eventually to the

"Space Olympic Games" in which special zero gravity competition complements the usual Earth-based sports.

Movie studios will be able to film a great number of special effects with complete realism in orbiting studios at lower cost than setting up elaborate sets on the ground. Advertising studios will be created in space in these and other specialized facilities in which ads will be created for consumption on the ground as well in space. Specialized manufacturing and processing facilities will manufacture products that take advantage of the zero or partial gravity environment, process material brought from the moon or other celestial bodies, and make finished products for export to Earth as well as to space stations and nascent lunar bases. In fact it is likely that a number of related activities such as the above will be aggregated in "space business parks" that feature common support subsystems, privately owned and leased to specialized commercial tenants as well as to some government users. Thus space businesses driven by the entrepreneurial spirit will flourish in a manner not different from myriad businesses on the ground.

These nontraditional commercial uses of space will become economically viable principally because of the large reductions that will occur in the cost of access to space. These cost reductions will occur even without matching decreases in the cost of spacecraft or space operations, and without the advent of new materials. However, the number and magnitude of such commercial space applications will literally explode with the major reductions in spacecraft and launch vehicle weights and costs that will take place once Buckytubes are used for manufacturing most structures and components of spacecraft and launch vehicles.

Estimates of the annual total commercial mass placed in space from all potential commercial activities in orbit by 2030 identified to date, without the introduction of Buckytubes, are 100 million kilograms—which is about 100 times the current annual worldwide mass placed into orbit. The annual mass is estimated to reduce with the incorporation of Buckytubes, but will still be about 30 million kilograms by 2030 because the wholesale adoption of Buckytubes will not be realized until the latter portions of this time frame. These are staggering numbers and indicative of the huge profit potential for space entrepreneurs, which is the incentive that will essentially guarantee that such businesses will be created in space.

It is for the above reasons that commercial ventures will dominate all uses of space and will constitute the second industrial revolution, this time in space. This "space industrial revolution" will fulfill the final perspective listed in Table 19.1: It is far better to create wealth via new ventures and enterprises, than to manage scarcity by tweaking the margins of current ones in order to minimize budget increases. The new wealth created will not stay in space, but will spread throughout Earth-based economies and ultimately benefit all Earth's people.

21 Final Thoughts

The preceding chapters have introduced a number of new technologies and techniques, and many new space system application concepts that they enable, to illustrate future space capabilities. It is clear that uses of space in the long term future (20–30+ years) will be very different both from what exists today or from what can be visualized by linear extrapolation of today's technologies and techniques.

Space systems will be at the same time enormously powerful yet much more affordable. Space operations will be no more expensive nor difficult than ground or most aircraft operations today. Defense uses of space will proliferate, and space will play a fundamentally different role in military functions than it does at present. NASA's long-term objectives will be attainable sooner and for a fraction of the budgets envisioned currently. However, the most profound change will be the emergence of wholesale commercial uses of space, whose sheer volume and breadth of functions will swamp all government uses of space.

This massive commercial activity in space will provide goods and services to Earth-based as well as to space-based customers, both private and government. These commercial activities will represent the long-sought large markets that are needed in order for large privately financed space ventures to flourish with no government support other than to prove the advanced technologies. We will finally see an end to the "Catch 22" snare: "Can't get the funds to develop the technology because a sufficient market does not exist, yet the applications that would constitute the market do not exist because the technology is not developed."

Some final thoughts as a result of the foregoing are outlined in Table 21.1.

These achievements will be the result of developing new technologies and techniques and applying them in new ways, which will create new capabilities that are neither achievable with current technologies nor by their linear extrapolation. We must set hugely ambitious goals and then seek the aptly called "disruptive technologies" that engender paradigm shifts to reach them. This means, in effect, that "you

Table 21.1. Final Thoughts

- Technology will make the space future very different from today.
- Systems will be enormously more capable, yet very much lighter and cheaper.
- Extrapolation of today's techniques won't get us there.
- We must be willing to think big, far term, and high risk (even though this is contrary to most current practice).
- Otherwise we will miss the boat.
- Others will implement the new space systems, whether we do or not.
- Space will finally become just another place.
- The specific applications shown are less important than these points.

can't get there from here by a straight line." To attain this desirable goal we must be willing to think unconventionally, big, far term, and high risk. This is as "out of the box" as it gets. The problem is, as ever, that such thinking is contrary to most current practice, which focuses on, and rewards, incremental evolution.

It is imperative that those advocating "disruptive technologies" (and they do exist) be listened to and their ideas be embraced wholeheartedly, or we will miss the boat. Make no mistake, advanced space system concepts such as those discussed in this book will be implemented by others, whether we do or not. This is because the technologies will be known to all, and the enormous system capabilities that will be attainable for relatively minuscule budgets will make them attractive to, and attainable by, smaller economies including rogue states and even some terrorist groups. If we forfeit these space capabilities, we leave the arena open to others to dominate space inexpensively for profit, prestige, power, or ideology. Is that the future we want?

It has been the dream of space planners for many years that "space become just another place." This is because space has always been a "special" place, principally because it is so difficult and expensive to attain and systems in space are essentially unreachable. With such thinking, space systems must be super reliable and simultaneously extremely lightweight, which are the opposite characteristics of normal activities on the ground. Thus "special" was and remains an essentially negative appellation.

When the difficulty and cost of development, launch, operations, and maintenance of space systems reduces to the point of becoming comparable to those of terrestrial commerce, or even airborne operations, space will finally cease to be special, and become just another place for routine activities and business. The "places" will then expand from "ground, air, or sea" and become "ground, air, sea, or space." This will be as it should be. It is the author's fervent hope that the technologies, techniques, and space system application concepts presented and discussed in this book will help to make that come true.

Two repeat caveats are mandatory. The first is that the space system concepts discussed in this book are illustrative examples of what might be done with new technologies and imagination and are not, in any sense, a complete set. Other exciting concepts surely already exist, and many more will undoubtedly be advanced by others. The second caveat is that attractive as some of these concepts may appear, this book does not advocate them either individually, in a subset, or the entire set. This is because responsible advocacy must take into account quantitative comparisons with other means to accomplish the same objectives as well as broader national considerations, neither of which were treated in this book.

In the above context it must lastly be pointed out that the specific system concepts presented, whether individually or collectively, are less important than the major observations, perspectives, implications for the space program, and these final offered thoughts.

Sources

Part of the content and approach of this book is based on the 1976 Aerospace Corporation study for NASA on advanced concepts that culminated in a report entitled "Advanced Space System Concepts and Their Orbital Support Needs 1980–2000," The Aerospace Corporation Report No. ATR-76(7365)-1. I directed the study and wrote the final report with H. L. Mayer and M. G. Wolfe. In 1997 The Aerospace Corporation asked me to update that original study to project the future scenario out to 2030 and beyond, which resulted in the report "Advanced Space System Concepts and Enabling technologies for the 2000-2030 time period," Revision 1, July 7, 1998, by Ivan Bekey, Bekey Designs, Inc. for Aerospace, Contract No. 46-00000-571. That study was commissioned by Dr. George Paulikas, who at the time was Executive Vice President of Aerospace. I performed that study in its entirety under the general guidance of Dr. Larry Greenberg, general manager of laboratory operations at Aerospace El Segundo Offices, with the notable encouragement and support from Drs. Vincent Boles, Laurie Henrikson, and Grant Aufderhaar of the Advanced Technology Division of the Aerospace offices at Chantilly, Virginia.

Most of the concepts and the ideas for new technologies in this book originate with those two studies and were taken from those two reports. These were augmented by other concepts and technologies to which I was exposed in my 48 years of professional experience, including 18 years at Aerospace, 19 years at NASA Headquarters, three years at RCA, and three years at Douglas Aircraft, as well as countless discussions in workshops, symposia, and conversations with many friends and colleagues from government, industry, and academia. Because these concepts are by definition advanced, their description and characteristics evolved as they matured and frequently inspired me to create improvements or wholly new application concepts, as the impetus for these inventions was always to seek useful concepts to address real world needs. Thus some ideas in this book are variations of previously known concepts, while others I conceived in the process and are partially or wholly new.

Notwithstanding the above I would like to acknowledge a number of individuals who were particularly instrumental in my gathering sufficient information to formulate and write the reports and the book. This process is fraught with difficulty, however, because many of the concepts have had a long history of innovation, refinement, dissemination, and advocacy by a number of people, and it is difficult to give credit for the ideas to any one person or even a group of persons or company without seriously slighting others. I would like therefore to give a few examples of the history of conceptual development of some the ideas to illustrate this quandary.

A typical concept with a long evolutionary history is that of delivering energy from space to the Earth—the space solar power concepts presented in Chapter 7 on energy beaming. The space solar power delivery concept had its origins in Dr.

Peter Glaser's conception as described in his article "Power from the Sun: its Future," in *Science*, vol. 162, pages 857-866 (1968); and his tireless advocacy of the concept as a means of delivering inexpensive, inexhaustible, and nonpolluting power to developing and industrialized nations alike. Countless papers have been written on the subject, and indeed continue to be presented, witness G. A. Landis, "Reinventing the Solar Power Satellite," NASA Glenn Research Center, Cleveland, Ohio. Paper No. IAC-02-R.1.07 presented at the World Space Congress, Houston, Texas, October 10–19, 2002.

NASA and the Department of Energy jointly performed extensive system design and implementation studies and experiments in the 1970s to define the concept and assess its programmatics including costs, and issued a major report "Solar Power Satellite System Definition Study," Document D180-25641-2, NASA/DOE, November 1979. The National Academy of Sciences was then asked to assess the concept and issued a report "Electric Power from Orbit, a Critique of a Satellite Power System," National Research Council, July 1981. The fairly negative tone of this assessment caused the concept to lay dormant for many years because it could not compete economically with conventional means of power generation on the ground and because it required an enormous initial investment.

I included the concept in the 1976 study report because of its great potential, and it continued to find its way into professional papers given and published for the next 20 years. In 1995, while Director of Advanced Concepts at NASA Headquarters, I commissioned a new look at the space power concept to reassess the potential of the idea using technologies that had emerged since the original studies, alternative architectures, and extensive use of robotics rather than astronauts. The study was led by John Mankins at NASA, and resulted in a major NASA/contractor study report known as the NASA "Fresh Look Study" ("Space Solar Power: A Fresh Look At The Feasibility Of Generating Solar Power in Space for Use on Earth," NASA Report No. SAIC-97/1005 to NASA HQ, April 4, 1997). The study showed that by using advanced technologies, new architectures, and approaches better suited to the space environment, the system could indeed be made much more economically viable and competitive.

I included two implementations of the solar power satellite concepts in the 1998 update of the 1976 Aerospace study and in Chapter 17 of this book. The concept continues to evolve partly because of NASA's recent funded study activities and partly because of the increasing interest of other nations as expressed in professional symposia. Thus, while my two examples of the concept may not represent the absolute latest in concept development (which is beginning to take a serious look at lasers for power transmission and more advanced component technologies), they will serve to illustrate the ideas at a reasonably mature level. However, because of the involvement of so many people from industry, NASA, and academia, it is not possible to credit a particular individual or even a few individuals with the concept as currently understood without slighting many others equally deserving.

Another concept, whose history illustrates this difficulty, is the tether application concept in Chapter 5, which used momentum transfer in a rotating tethered system in space to transfer energy and angular momentum among the end masses. This concept originated under my guidance at NASA, where I was the principal sponsor and advocate for applying tethers in space. The concept was actually conceived within a NASA Tether Applications Group that I chartered for a number of years beginning in the 1980s, and whose product was a large number of applications for tethers in space, including this concept, as described in the NASA publication *Tether Applications Handbook,* 3rd edition (December, 1997), edited by M. L. Cosmo and E. C. Lorenzini, Smithsonian Astrophysical Observatory, NASA Marshall Space Flight Center Grant NAG8-1160. This publication is also available for downloading on the NASA MSFC web site.

In 1996 I performed a study to define the concept further in specific application to increasing the payload that a given launch vehicle could launch into space, and to quantify its advantages compared with conventional means to launch payloads. I was supported in this early study by Joe Carroll of Tether Applications and by Ted Talay of NASA Langley Research Center. The findings were published in an unnumbered NASA memorandum that was circulated at NASA and to industry, academia, and congressional staffs. The following year I published the concept and results of the calculations in a paper "Increasing the Payload and Decreasing the Cost/Pound of an RLV by Using a Tether to Deploy the Payload," which I presented at the International Astronautics Congress (Oct. 1997) (IAF-97-V.5.09). The concept remains a viable option for near term application, but attempts at advocacy have not met with success. This concept is another example of the long and complex background of development of a particular idea which, as a result principally of its early history, makes it difficult to give credit to any single source.

Yet another kind of concept history, in which there were a relatively small number of contributing individuals and who can therefore be identified, is the concept for using spectrally matched multiple bandgap cells inserted into the rainbow created by an optically dispersive element such as a prism, and also using concentrating films to attain very high efficiency solar power. This concept, discussed first as a general technological principle in Chapter 2 and again as related to specific concepts in Chapter 6, was hinted at in diverse papers and discussions in the 1950s and 1960s, two of which were E. D. Jackson, "Areas for Improvement of the Semiconductor Solar Energy Converter," *Transactions of the Conference on the Use of Solar Energy,* University of Arizona Press, Tucson, AZ, Vol. 5, pages 122–126 (1955); and in the ad-hoc panel on solar cell efficiency, Space Science Board, National Academy of Sciences, Washington, D.C., "Solar Cells, Outlook For Improved Efficiency" (1972). Dr. Wade Blocker of Aerospace next conceived an implementation of the spectrally split approach to high efficiency and published

"Projections For Spacecraft Power Supplies in the Post-1985 Time Period," The Aerospace Corporation Report No. TOR-0076(6020-01)-2 (February 1976).

Dr. Blocker gave a large number of presentations to government and industry people on this concept. He also wrote a paper that was published as "High Efficiency Solar Energy Conversion Through Flux Concentration and Spectrum Splitting," *Proceedings of the IEEE*, Vol. 66, No. 1, pages 104–105 (January 1978). At about the same time I proposed to NASA that Aerospace perform a study to address the means to implement the concept and project realistically attainable performance. NASA contracted the study, which I directed and for which I wrote the final report: "High Efficiency Solar Photovoltaic Power Module Concept" The Aerospace Corporation Report no. ATR-78(7666)-1 (March 1978). The concept apparently then lay dormant for two decades. Subsequently while at NASA Headquarters in the late 1990s, I commissioned and funded an activity at the Jet Propulsion Laboratory (JPL) to perform laboratory experiments to attempt to experimentally verify the potential of the spectrally split, concentrated, bandgap matched solar cell technique. This work resulted in a number of NASA presentations and professional papers intended to elicit public interest and involvement, and further its potential commercialization.

This activity was directed by Dr. Neville Marzwell at JPL. The first professional paper on this subject was presented at the 32nd IECEC conference in Hawaii, July 27–August 1, 1998, and published as: "Multi-Band Gap High Efficiency Converter (Rainbow)," (paper no. 97228) by Carol R. Lewis, Wayne M. Phillips, Vigil B. Shields, Paul M. Stella, and Ivan Bekey. The concept was further disseminated as a paper and presentation at the 1998 49th International Astronautic Congress, Melbourne, Australia September 28- October 2, 1998: "High Efficiency Space Solar Power Conversion System," (paper no. IAF-98-R.3.08) by N. Marzwell and W. Phillips. I later published an article describing the concept and its performance results in the American Institute of Aeronautics and Astronautics monthly magazine *Aerospace America* entitled "Rainbow's Array Of Promises" (January 1999).

The concept was reduced to practice in laboratory tests under the general direction of Dr. Marzwell and is now considerably more refined than in 1999. Since the individuals who made most of the contributions to that concept are fairly few, they can be clearly identified for source credit. They are E. D. Jackson, Wade Blocker, Ivan Bekey, Neville Marzwell, Carol Lewis, Wayne M. Phillips, Virgil Shields, and Paul Stella. There are probably others who have continued to mature the concept since I wrote this manuscript, but one must close the writing sometime or the text is never finished.

A different history applies for the concept for "Earth-Stationary Pole-Sitter Spacecraft" in Chapter 16, which was conceived and actually patented more than 25 years ago by Dr. Robert L. Forward, recently deceased. Dr. Forward never hesitated to include the description of the concept and its operation in his very many public lectures. Robert and I were good friends and frequently shared new ideas and technical discussions, and he discussed this concept with me with no reservations or limitations. Since in this case there was clearly a single source for the concept he deserves full credit as the originator.

The source process was yet again different for several concepts evolved most recently for government agencies, including the large aperture optical concepts described in Sections 15.20–15.25. The heart of these concepts is the adaptive piezoelectric membrane. The concept was approached by a number of investigators including J. Feinleib, et al., "Monolithic Piezoelectric Mirror for Wavefront Correction," *Applied Physics Letters*, Vol. 25, No. 5, pages 311–313 (September 1, 1974). Experiments to demonstrate the technology were proposed by John Main, University of Kentucky, and funded in 1996 under my Advanced Concepts Research Program at NASA Headquarters, which was chartered to produce material for public release. This resulted in a number of papers by Main, including "Control of Distributed Actuators Without Electrodes," *Proceedings of the International Engineering Congress and Exposition* (Dallas, Texas, November 14-18, 1997). A year after I retired from NASA, I proposed to the NASA Institute for Advanced Concepts (NIAC) a Phase I study to apply the adaptive membrane concept in combination with my ideas on precision formation-flown elements to replace large structural trusses in space to a large space telescope. The resulting report indicated feasibility and was published on NIAC's web site for public access (www.niac.usra.edu). Contributing to the report were Glenn Zeiders of Sirius Group, Fred Hadaegh and Edward Mettler of JPL, Nelson Tabirian of BEAM Engineering, and Samuel McWaters, Kevin Bell, Siegfried Janson, and Richard Boucher of Aerospace.

Subsequently I presented a paper on the subject at the AIAA Conference on Space Systems and Technologies, Albuquerque, New Mexico: "A 25 Meter Space Telescope Weighing Less Than 100 kg" (Ivan Bekey, Bekey Designs, Inc., paper # 99-4478, September 29, 1999). I then investigated the concept in more depth for the government as applied to high performance ground imaging. I was the Principal Investigator, and was supported by a subcontractor team represented by Nelson Tabirian of BEAM Engineering, Ed Mettler of JPL, Enrico Lorenzini of the Smithsonian Astrophysical Observatory, Jim Harvey of University of Central Florida, Leo Lichodziewski of L'Garde, and Eric Austin and Dan Inman of Virginia Tech University. I presented a paper summarizing this work at the International Society for Optical Engineering (SPIE) Astronomical Telescopes and Instrumentation Conference "Very Large Yet Extremely Lightweight Space Imaging Systems" (Ivan Bekey, # 3543, August 22-28, 2002, Hawaii, USA).

Many of the concepts evolved for the U.S. Air Force in the 1970s were done so under classification restrictions that prevent me from citing their many early contributors. However some of these concepts were formally declassified on May 6, 1998, by the Air Force Space and Missile Systems Center, and thus some mention of their principal contributors is possible. As an example consider the concepts 15.7 and 15.16, dealing with detecting and communicating with submerged submarines. These concepts were originated by Dr. Harris Mayer and myself while working collaboratively at Aerospace on the 1976 Advanced Concepts Report. Wade Blocker, Rudi Meyer, George Paulikas, and others contributed to the concept development, but the bulk of the calculations were done by Harris Mayer and the implementation of the large photon bucket in space was my conception. The concept was refined after numerous presentations to various levels of the Air Force and DOD during which adjustments were made to make the concept more relevant to needed functions.

An example of technology conceptual development pertains to the large space-fed multibeam antennas that I used in many of the radio frequency concepts. I developed the concept of a large space-fed array antenna in the early 1970s while researching the most feasible way to implement a large RF aperture capable of electronic beam steering and multiple simultaneous beam simultaneously. This was needed to enable a number of concepts where the antenna had to be in GEO for continuous coverage and had to be extremely large in order to have the required aperture to detect signals from weak sources on the ground, but its very small footprint required beam replication in order to obtain large coverage simultaneously as well as the frequency reuse to become practical given spectrum availability limits. Supporting the conceptual development of this antenna concept and calculation of its predicted performance was principally Leo Cantafio of Aerospace. Calculations of the communications link budgets were supported principally by Herb Wintroub of Aerospace. While the concept was developed primarily to support the feasibility of the personal communications concept using "Dick Tracy Wrist Radios" described in Section 14.11, the antenna implementation was used with parameter changes in many of my concepts that appear in Chapter 14—Concepts Employing Radio Frequencies and Microwaves. Most instrumental in the acceptance and dissemination of these concepts were Pete Leonard and Sam Tennant of Aerospace and Bob Freitag and Phil Culbertson at NASA Headquarters.

The foregoing examples serve to point out the complex origins of most of the concepts in this book. Because of this, and also because I do have even a reasonably complete history of the development of most of the concepts, in general I have not cited sources and individual contributions to the concepts in the text; though in a few cases I was able to cite sources or acknowledge individual contributions in the text pertaining to the idea.

Nevertheless it is possible, and very appropriate, for me to mention some people who have helped me to understand the technologies and principles involved in a number of key discussions, and I would like to acknowledge them here, recognizing that I probably missed some good people, whose forgiveness I hereby beg.

Contributing significantly to the 1998 report were the following persons, listed under the relevant sections in the book.

Technologies

Siegfried Janson, Richard Boucher, and Harris Mayer, of Aerospace; Neville Marzwell, Dick Dickinson and Randy Bartman, of JPL; Joe Carroll of Tether

Applications; Robert Forward of Tethers Unlimited; Rick Smalley and Ken Smith of Rice University; Patrick Collins of NASDA; and others.

Radio Frequency

Some of the above people, plus: Mario Grossi, Smithsonian Astrophysical Observatory; Wayne Sievers, MITRE; and several people at Lincoln Laboratory, RAND, and other organizations.

Optics

Some of the above people, plus: Krafft Ehricke of Rockwell International; Rod Hyde and Lowell Wood of Lawrence Livermore National Laboratory; Howard MacEwen of the National Reconnaissance Office; and several people at the Air Force Research Laboratories; and others.

Transportation and Infrastructure, and Energy Delivery

Some of the above people plus: John Mankins, John Anderson, and John Rather of NASA HQ; Dennis Wilson of the University of Texas; Doug Stanley and Dennis Bushnell of NASA Langley Research Center; Mike Yarymovich of Rockwell International; Paul Penzo of JPL; Leik Myrabo of Rensselear Polytechnic University; Jim Powell, Energy Sciences; and others.

In addition, I would like to note the contributions from a number of individuals who were consulted during the initial 1976 study who were acknowledged in that report for their "significant impact on the perspectives of how to operate in space, on types of initiatives which might be meaningful, on requirements, or on all the foregoing." They were: Bob Cooper, Dan Brockway, Howard Barfield, and Jim Wade from the Office of the Director of Defense Research and Engineering; Joel Bengston and Rex Finke of the Institute for Defense Analyses; Jasper Welch of the Office of the Secretary of Defense; Gerry Dineen and Dan Dustin of MIT Lincoln Laboratory; Dick Garwin of IBM; Gerry Sears, Russ Sharpe, Joe Mate, and Ted Parker of RAND Corporation; Ernie Martinelli and Dr. Bhatachari of RDA, Inc; Freeman Dyson, Mal Ruderman, Steve Weinberg, and Ken Watson of the Jason Group; Ben Alexander, John Ise, and Dick Holbrook of General Research Corporation; and Milt Birnbaum, Tom Hartwick, and Tom Taylor of Aerospace.